Praise for

TEN
SURVIVAL
SKILLS
FOR A
WORLD
IN FLUX

'Original and thought-provoking: a manifesto for a better way
of educating humanity' GORDON BROWN

'Challenging and hopeful: a groundbreaking guide to the
future' VALERIE AMOS

'A fascinating and inspiring analysis of how the world is
changing and education needs to keep up'
RACHEL SYLVESTER, *The Times*

'Brilliant, an absolute must-read. A riveting, superbly written
account of the world today, and tomorrow. One of the books
of the year from the Renaissance Man of Oxford'
MATTHEW D'ANCONA, author of *Post-Truth*

'While Fletcher structures his analysis around 10 skills, his hints, tips, advice, comments and things-you-and-I-can-do-today to bring about change are so frequent that I lost count. Not 10 ideas, not even 57, but countless. I made notes of quotes from what is a provocative and hugely thoughtful compendium of positive and realistic thinking to navigate an increasingly difficult world. It's leavened by the wisdom of dozens of prominent thinkers and doers, many of whom Fletcher has met personally. The positive tone is admirable and ultimately infectious but only because Fletcher does not dodge the profound difficulties facing us now as individuals, nations, governments, businesses and as a planet ... offers a structure for seeing, naming and ultimately dealing with the torrent of problems now and to come. His account is enlivened by the thoughts of those he's worked with – including intellectuals and thinkers from around the globe, and some words with world leaders he happened to meet, including Barack Obama ... Fletcher's core message is, for me at least, realistically optimistic. For humanity to survive and thrive we must be both curious and co-operative'

GAVIN ESLER, *The National*

'Tom's pot pourri of insights and practical ideas gives us a pathway for the future we have yet to live, in a world that can be hard to navigate'

CATHERINE ASHTON, former EU Vice President

'A magnificent book! As we (anxiously) anticipate the rest of this century, Tom Fletcher presents us with a brilliantly timed, incredibly important, gem of a book – overflowing with wisdom and much hope too'

ZEID RA'AD, former UN Commissioner for Human Rights

'An excellent handbook on how to coexist not just with each other, but with technology too'

MUSTAFA SULEYMAN, Founder of Google DeepMind

'Survival forces one to avoid disaster by action, complacency lulls one to its inevitable outcome. This manual for humanity is a call to action for all of us. Read it, learn from it, act aggressively and uncompromisingly upon it, let it guide the rest of your life's work. Maybe then hope can be rekindled. If you disagree then feel free to reflect in less than a decade on just how wrong you were, before the lights turn off'

GENERAL SIR GRAEME LAMB, former SAS commander

'Global education has found a champion. Future generations have found a spokesperson' ANDREAS SCHLEICHER, OECD

'A scintillating humanifesto for creativity in how we learn. Inspiration not just for parents and teachers, but all of us'

ANDRIA ZAFIRAKOU, Global Teacher of the Year 2018 and author of *Those Who Can, Teach*

'The international community needed a call to action. This is a must read for the inventors, dreamers and pioneers of our future'

DUBAI ABULHOUL, Young Arab of the Year 2016 and Founder of the Fiker Institute

ALSO BY TOM FLETCHER

The Naked Diplomat

TEN SURVIVAL SKILLS FOR A WORLD IN FLUX

TOM FLETCHER

WILLIAM
COLLINS

William Collins
An imprint of HarperCollins*Publishers*
1 London Bridge Street
London SE1 9GF

WilliamCollinsBooks.com

HarperCollinsPublishers
Macken House, 39/40 Mayor Street Upper
Dublin 1, D01 C9W8, Ireland

First published in Great Britain in 2022 by William Collins
This William Collins paperback edition published in 2023

1

A catalogue record for this book is
available from the British Library

ISBN 978-0-00-844791-5

Set in Sabon LT Pro
Printed and bound in the UK using 100%
renewable electricity at CPI Group (UK) Ltd

To my parents, Mark and Debbie.
Education pioneers.
And kind, curious, brave ancestors.

Contents

PART TWO: Begin It Now

Prologue

Zeinab's Moonshot

This journey started with a question that I failed to answer.

Lebanon's Bekaa Valley is one of the most beautiful places on earth. For centuries, empires have fought to exploit its fertility and control its location at the hinge between continents. Visitors have sunbathed while looking up at snow-capped mountains. Famously the Roman Empire survived on bread and circuses. The bread came from here. So did much of the wine for the circuses. The awe-inspiring ruins of the temples of Bacchus and Jupiter at Baalbek in the Bekaa Valley are testimony to the value to Roman civilisation of both.

More recently this place became a centre for the cannabis cultivation that helped fund a devastating Lebanese civil war. The tourists have since dried up, deterred by travel warnings and the yellow flags and roadblocks of Hizballah. In the Bekaa Valley, it is refugees who now dominate, the battered human overflow from conflicts in Syria, Israel/Palestine and beyond.

Our convoy had bumped along the dirt road in the evening drizzle. This was the last stop of a long day touring the bedraggled refugee camps that were now home to Syria's battered civilians.

I met Zeinab outside the small tent she shared with her extended family. The shelter had a roof made of old tobacco advertising boards. Twelve years old, she told me she had been out of formal education for four years. Her family had fled Homs, a town now pummelled to rubble by the barrel bombs of Syrian president Bashar al-Assad. Her school, home, hospital and innocence had been destroyed. With her father killed and her mother working long hours in a bakery outside the camp in which she now lived, she was bringing up her two younger siblings, including trying to teach them to read and write.

Zeinab gestured to a corner of the tent. Her youngest brother, Ahmed, was kneeling on the damp floor, drawing with a stubby black crayon. He was my son's age, eight, but malnutrition and a wariness of adults made him seem much younger. Hesitatingly, he showed me the picture: stick children and war planes. The planes were shooting stick missiles at the stick children. The stick children were either running or lying on the ground.

Food and clean water were scarce in the camp, Zeinab said. Medicine even more so. But that was not what she wanted to talk to a visiting ambassador about. Her light-brown eyes were shining at the possibility that a makeshift school might soon open up in the neighbouring camp.

'I want to be an astronaut,' she said.

This was not what I had expected. But it was good to have a diversion, amid the gloom. 'That's brilliant, do you want to go to the moon?' I mimed a rocket, fearing my Arabic might need some reinforcement.

'No,' she replied in accented but perfect English. 'I want to find safer planets for Ahmed.'

I felt a wave of despair. Did I have any right to mouth the usual platitudes about dreaming big and working hard? She could do both and yet still be overwhelmed by the odds. Was it wrong to try to share any sense of hope when I knew that there was so little prospect of it being rewarded?

I stammered something about the need to keep going, to make time to study. She nodded gravely. This was clearly not the first time Zeinab had heard it from white men with photographers in motorcades, and it would not be the last. A bodyguard tapped his watch and flicked his eyes towards the cars. The Lebanese military had told us we could not plan on passing through Hizballah checkpoints after dark: an ambassador 'accidentally' taken out in the conflict zone in which we were travelling would have exactly the chilling effect on Western engagement that the regime in Damascus sought.

Perhaps somewhere a teacher might battle exhaustion and class numbers to help Zeinab and Ahmed catch up. An NGO campaign might shift more funding from all the other life-saving priorities towards education. A Western politician might face down the corrosive drumbeat of tabloid racism and give them a chance to start again in a new country. Luck, resilience and kindness might conspire to help them defy the odds, and maybe Zeinab would get her moonshot?

But it would take a massive amount of luck, resilience and kindness.

As I turned to leave, Zeinab grabbed my sleeve. Her face was earnest. 'But what do I really need to learn?' she asked.

What are the essential building blocks of knowledge, the ideas and values we are most fortunate to inherit from our parents, teachers and ancestors? How do we ensure that our children are prepared for a world that we cannot imagine? What do they need to know to live a good life? Each of us is a bridge, what my father calls the story bearer, between the past

3

and the future. It is a daunting responsibility, and one for which we don't have a guidebook.

I drove away from the camp frustrated at my failure to articulate how hard these questions are for all of us, however privileged our lives are compared to those of Zeinab and Ahmed. But more importantly, furious that I had not been able to help her imagine a world in which the rockets her brother drew were aimed at the stars and not at them.

As she ducked inside the tent, Zeinab had turned back, a thoughtful grin playing at the edges of her mouth.

'I will do what I have to do. Will you?'

This book is my attempt to give Zeinab a better answer.

Introduction

Kindling the Flame

Education is the most powerful weapon which
you can use to change the world.

Nelson Mandela

This book is not about how you can be thinner, richer and
smarter, though it might just help. It is not about how you or
your kids can get ahead, though it might just help with that
too.

Instead, this is a book about survival. As individuals, families, communities, society. It is about what we need to learn if
we are to find better ways to live together, better ways to
respond to the threats and opportunities we face. At a moment
of flux, when so many of us feel uncertain, and when the
ground beneath us feels less stable, it might help you feel a
little more in control. My hope is that this is also a book that
helps us become better ancestors.

After meeting Zeinab and Ahmed, I left diplomacy to work
to get education to the million Syrian kids who are out of

school. I argued with governments and business leaders that we faced a simple choice: whether young refugees came to us in doctor's jackets and medical vests or life jackets and – in some cases – suicide vests.

In the course of that work, time and again young people and their parents told me that what they wanted most was education. Yet as they moved between countries seeking safety, they also moved through multiple education systems, losing valuable time and vital hope. I met one Afghan refugee, Khaled, who had been part of five different education systems in a year. Migrants in Europe are twice as likely to drop out of school. More widely, six out of ten young people across the world can't read or add up, and a staggering seventy-five million are not receiving any formal education at all.

But Zeinab's question is a much broader challenge than simply getting more kids into school. What do they really need to learn when they get there? And what do the rest of us need to learn that we missed when we were at school ourselves? I spent two years at New York University leading a team that tried to answer Zeinab's question. How can we ensure that humans learn the right things in the right way? What are the new survival skills, and how do we develop them? I interviewed hundreds of people, from special forces commanders and refugees to prime ministers and pioneering educators to adventurers and tech titans.

More importantly, with Zeinab's question burning, I set up student hackathons in Abu Dhabi, Shanghai, Sydney, Madrid, Nairobi and New York – intense, collaborative events at which young people were thrown together and faced with a challenge. I asked them to identify what they weren't being taught yet needed to know. The process was deliberately chaotic, to allow space for students to be creative. Many of my colleagues thought I had finally lost it. But as the project grew it was clear that the students were relishing the chance to think about what they were missing. I had expected an exhausting laundry list of ideas. However, as the responses came in, wherever they were from and regardless of their educational background, these young people had a consistent sense of what they wanted.

What did they tell us should be part of a curriculum in and for the twenty-first century?

'Where did we come from, and how we did we get here?' They were aware of a sense of shared human identity and history, yet baffled that they didn't study global history. They wanted to know how their country's story seemed to others, and how it linked to the stories of other countries. 'We don't want to study history as us and them, as winners and losers.'

They felt that they needed a basic grounding in politics, civics and ethics. 'What are our obligations to each other?' 'What are the limits on our freedoms, and how do these change from country to country?' 'How have humans governed themselves through history?'

Strikingly, all groups felt there was a great gap in their understanding of their health, physical and mental. They wanted to learn emergency first aid, to get diet and exercise advice, sexual education. While not especially glamorous or revolutionary, simple life skills such as finance, cooking or vehicle maintenance were important to them too. 'Maybe previous generations learned these practical skills from their family or community?'

They consistently raised the need for digital skills, not because they would necessarily use them, but because they are a language to master. 'We won't all become coders, but we need to make sense of the technology around us.'

They felt that education systems did not give them enough guidance on how to articulate opinions, build arguments and influence others. They wanted to learn to listen properly, and engage and debate with others constructively, rather than just project their own voices. They wanted to develop tolerance, respect and openness towards different groups of people, as well as empathy towards others. They stressed that this didn't just mean respect for people from a different geographical or racial background. 'We need to know how to talk to people we disagree with.' They frequently spoke about resilience and the ability to deal with failure, citing the anxiety and pressure of their lives. 'We are always on, never off.' They regularly articulated this in the context of a more polarised debate. 'How can we find our voices, and know when to speak?' 'How do we know when we are being misled?'

'How can we best learn?' They had a very varied experience of learning, from minimal education to factory schools, in which they sit in rows of neat tables, to elite independent schools. But consistently they made the case for more autonomy and flexibility. They wanted more learning outside the classroom, more music and sport. They wanted to travel more, to mix with people from other cultures.

They were frustrated that academic skills seemed to be more important to parents and teachers, and that assessments were weighted towards exams and memorising facts. 'We should be tested on our ability to work in groups and teams, to crack problems and do projects together. We should be encouraged to take risks, not discouraged.'

Most felt that technology was not yet being used to its potential. 'This could give us access to so much more, but we're always told to switch off our phones. We can't access the internet during exams, and yet that is what everyone does all the time in real life.'

The hackathons confirmed to me the deep desire for change, and the urgency of renewal of our education systems. Most notably, the exercise drove home the importance of involving learners in designing their own learning. As the philosopher Xun Kuang put it, 'Show me and I forget. Teach me and I remember. Involve me and I learn.'

I was provoked, challenged and inspired by what I heard. And by the questions they asked. Can we get better at living together? Why are we losing that vital connection with what Abraham Lincoln called 'the better angels of our nature'? Why do technology and social media seem to be encouraging too many of our worst instincts and too few of our best? Can we reclaim the space for compromise, understanding and reason? Or will the first prototypes of a global society be shaped by autocrats, extremists and those who believe that the answer to the twenty-first century is to build a bigger wall?

With the help of these young people, I turned their feedback into a course on twenty-first-century survival skills, which became this book. It attempts to fill in those gaps. Through doing that, I hope it will also give Zeinab a better answer to her question.

We navigated previous periods of peril and change through not just mastery of new tools but mastery of a new mindset. We cannot deal with the threats to our survival without doing that afresh. If a generation of humans online and on the move is not equipped with the skills they need, then extremism, inequality, drift, intolerance and distrust will increase. If we fail at this, our children and grandchildren will find themselves the refugees of the near future.

The Covid lockdowns of 2020–21 have driven us further apart physically, forcing us to think about what human connection we value, and what we can survive without. But they should also galvanise us to focus more time and energy on what we *cannot* survive without. We can – and must – be better prepared for what lies ahead. Through our research at

NYU, I concluded that humans will need a new balance between knowledge, skills and character. So the ten survival skills that this book sets out blend head, hand and heart.

For the head – knowledge – humans need a better understanding of the ingenuity and creativity that took us from cave painting to driverless cars. We will need to draw on the best of the knowledge that humankind has built over millennia. This will help us develop the curiosity to take the next great leaps forward. We will need a better understanding of how we learned to live together. Not just to study the periods of conflict, but to study the periods when we built societies able to reduce conflict. And we will need a more profound knowledge of our relationship with our planet. How can we get better at preparing for the future? What will the world look like in 2050, and what are the implications for our health, wealth and happiness? How have humans evolved, and how do we evolve afresh for a new set of threats and opportunities?

For the hand – skills – we will need to thrive, adapt, learn, create and coexist as global citizens. To get better at managing our physical and mental health. Classic curricula do not prioritise these basic living skills, which is why the bestseller lists are full of books about them. We need to build global competence: the cultural antennae that help us thrive amid different societies and cultures. We will need to learn how to learn, so that future generations will be adaptable enough to evolve in response to the technology tsunami ahead. As Harvard education professor Howard Gardner says, 'Don't ask how intelligent anyone is; but rather, how are they intelligent?'

For the heart – values – we must learn to be more kind, curious and brave. We will need to be kind enough to reduce inequality rather than widen it. We will need to be curious enough to invent new ways of living and organising ourselves.

And we will need to be brave enough to master technology rather than be mastered by it. How can we find purpose? How can we find, build and mobilise our tribe? How can we truly take back control? How can we coexist with technology and become better ancestors by tackling systemic injustices such as inequality, the climate crisis and inherited conflict?

Maybe, armed with these new – and not so new – survival skills, we can build networks in a time of institutional failure; consensus in a time of arguments; and bridges in a time of walls. We can strive for expertise, patience, perspective and judgement in a time of fake news, soundbites and echo chambers. We can aspire to be courageously calm, tolerant and honest in a time of outrage, intolerance and 'post-truth' politics. We can be internationalist in a time of nationalism, and open-minded in a time of closed minds.

That's what I hope my children learn. More importantly, I hope it is what the next Marie Curie, Albert Einstein, Al-Khwarizmi or Bill Gates learns. Perhaps our ancestors had worked out many of the answers to this challenge long before we industrialised learning. 'Education is not the filling of a vessel,' Socrates told his pupils, 'but the kindling of a flame.'

This transformation of why, how and what we learn will not be easy. But perhaps the stories shared in this book can help join some of the dots. Maybe the Oxford professor battling to create space on the curriculum for a global view of history can take heart from the art teacher striving to show that mastering creativity is not just an after-school painting club. The head teacher convincing teachers and parents that mindfulness helps academic success can take heart from the tech entrepreneur testing how play develops brain power. The business leader frustrated that their employees aren't equipped with the right problem-solving skills can take heart from the YouTube campaigner making popular videos on why education isn't working. The UN official exhausted by trying to make it easier for refugees to pass through multiple education systems can take heart from the students demanding they be taught global competencies rather than the list of wars their country won.

But why the urgency?

If we are to prepare ourselves and our children to survive the coming decades, our starting point has to be a better understanding of what actually lies ahead of us.

The question I get asked most regularly by students is: 'What job should I do?' The second is: 'Will it exist in thirty years?' No previous generation needed to be so aware of the precariousness of their chosen vocation or of the turbulence of a world in flux.

And we may be giving them bad advice.

Apply the same tests to almost any existing job and you find the same challenge: we fall back on what we know and understand. We spend more time with young people reflecting on the familiar pathway to a profession than if and why it will exist in 2050. When they are faced with such big, life-changing decisions as choosing a career, are we as confident as we should be in our advice? Should young people now really become lawyers, engineers, bankers, surgeons? Should they be learning what we learned?

What many school and university leavers in the West had in common in the year 2000 was that they didn't really think much about careers. Few really considered that they were signing up to a job for life, and yet the underlying assumption was often that they were. Many were looking for camaraderie, even if only in an office. Most aspired to a certain amount of prestige and money, but not huge amounts. Advances in social mobility made it less likely than for previous generations that they would enter the same profession as their parents. Yet it didn't feel precarious: they might not get their first choices, but something else would probably come up.

The decisions of school and university leavers were often driven as much by location (I want to travel, I want to be in London) as by a sense of craft, purpose or vocation. Few chose their jobs on the basis of how they could change the world. Yet the jobs quickly became a key part of how they defined themselves. They went straight from school or university to saying, 'I'm a mechanic/lawyer/doctor/teacher/accountant/diplomat.' Like our ancestors who literally took their name from their role in the community, the job was part of our identity.

The class of 2025 will look back on how many made their career choices in the late twentieth century in the way that we now think about the idea that we simply join the family trade. They will look back on the notion of a nine-to-five job with retirement at sixty-five like we look back on the idea that women don't work.

They are also leaving education with much more anxiety. With good reason. They will carry more debt. Getting on the housing ladder will seem remote, and pensions and retirement increasingly unlikely. They will more often have a sense that the first job they take will be a stepping stone rather than a life choice; and that they will move between jobs, not just over a lifetime but during their twenties. They will have a much better understanding that factors beyond their control – climate, economy, pandemics – will influence their destiny more than their choice of job.

This, and exposure to so much more information about the world, has given young people a sense of greater purpose and agency: they think more than previous generations did about making the world better. But it has also given them a greater sense of fragility. One in six young people in the UK have a probable mental health disorder. We are no longer taking for granted that our children's lives will inevitably be better than ours. Almost half of us think life will be worse. The figures are more optimistic in the East, but they are abysmal in Europe.

The class of 2025 will also already be thinking about more flexible career paths than their parents did. They will be less location specific: mobility matters more to them than place. They already want regular and Instagrammable experiences, not just an annual holiday fortnight getting sunburned on a beach. Realistically, they won't see owning a property as an early objective: as the *Economist* put it, 'Instead of houses, they will have houseplants.' As they hit the job market, they will find a very different terrain, with jobs more likely to include care worker, data detective and digital rehabilitation adviser.

They will be much more mobile. They will have a stronger sense of the distinctive skills they have and need. We will no longer operate in the twentieth-century model through which my generation moved – if we were fortunate – from being apprentices to toilers to experts to managers to partners to retirees. There will be even more intense competition from automation, offshoring and outsourcing. A single career will be very rare. The average time spent in a job is already only four years, and – according to the World Economic Forum – a third of the skills that workers need, regardless of their industry, will be obsolete by 2030. A quarter of adults already say that they don't even have the right skills for their current job.

For the first decade ever, most young people will also be digital natives. To take a twenty-two-year-old graduate in 2025: Skype was launched in the year they were born, Twitter was founded when they were three, and the first iPhone was released when they were four. Many young people in advanced economies have never lived or worked without this technology, and – during lockdowns – it was often all that connected them to their friends and universities. People of my generation frequently criticise young people for their relationship with their devices. But our research at NYU showed that they have a far better understanding, and control, of the technology – including social media – than we do. On that basis, it is children who should be telling their parents to put their devices down.

Work and life will blend further. Those leaving education now will be even more likely to have a diverse, even eclectic, portfolio of roles and interests than a single job. A quarter may have no full-time job, and some form of universal basic income will remain a distant prospect. On current projections, they will get married later and have fewer kids. It will be even more vital to have mentors; keep an open, learning mind; build networks; and look after their physical and mental well-being.

So much for 2025, but can we look further ahead to the children of those we are currently educating? The truth is we don't know what the world will be like in 2050. But we can make some educated guesses. Over pancakes in New York, Alec Ross, author of *The Industries of the Future*, told me the twenty-first century is a terrible time to have a single pathway: 'The employment landscape in 2050 will look nothing like today's. It is criminal negligence that we are not adapting our education systems to prepare for this.'

He's right. Almost half of today's jobs are at high risk. Classic office-based and manual roles will disappear by 2050, just as those jobs replaced the need to spin cotton, lift sacks or make horseshoes. Nancy Gleason, a professor of learning at NYU Abu Dhabi and author of *Higher Education in the Era of the Fourth Industrial Revolution*, told me that knowledge won't be enough for successful employment once machines take over pattern-based work. 'The well-paying jobs will involve creativity, data analytics and cyber security. As work becomes automated, it will also become much more fluid. Employees will need to be able to jump between very different types of tasks and contexts.'

Manual, repetitive tasks will be hit first by automation. Ask anyone who works or has worked as a switchboard operator or postal worker. But we can already see a wider set of jobs that are becoming vulnerable: taxi drivers, surveyors, fast food workers, translators. It is not hard to imagine how a satellite map and a chip in a car could replace the parking enforcement officer in the way that a speed camera has replaced its human equivalent. If you can't imagine how your industry or craft could be disrupted by technology, you will soon be hearing from someone who can.

It is a myth that those in more creative roles will be significantly better insulated from automation. We are not far off computers being able to generate more intelligent and coherent thoughts than most humans. Take GPT-3, a predictive language model that can code and write. It can't yet reason like a human, or pass the Turing test, convincing us that it is human. But the quality of its writing is the closest yet to ours, and – most importantly – it is improving much faster than we are. Creativity will be automated later than many other skills, but it is now possible not just to imagine but to experience a level of machine creativity that shows that our lead as humans is narrowing.

We don't know what the next phase of human evolution will encounter, but follow the money. Technology giants are investing heavily in robotics that will become integrated with our bodies to make us stronger, more resilient and smarter. As a result, the gulf between human and machine is closing: the entrepreneur Elon Musk has claimed that we are already cyborgs. If you were to walk down a busy street in 2030, you would be likely to pass people who have already had implanted a microchip for medical reasons or simply to reduce the hassle of carrying their keys and bank cards. If you were to enter the military, you would be likely to encounter people who are already using exoskeletons. If you work in diplomacy, you would be likely to talk to people using machine translation. If I am alive in 2050, I fully anticipate that I'll have artificial body parts that have been 3D printed.

Meanwhile, the Internet of Things gives to the machines around us an extraordinary ability to work together to monitor and organise our lives. Quantum computing and the artificial intelligence arms race are rapidly adding to the list of areas where human brains can't compete. And these challenges won't only confront the parts of the world reached by technology. Massive investment is going towards delivering the internet to the five billion people who don't yet have it.

These breathtaking advances are matched by those in genetics and nanotechnology. When Craig Venter, one of the first scientists to sequence the human genome, was asked if he was playing God, he allegedly replied: 'We're not playing.'

This is not science fiction. These waves of technological change are going to alter basic human capacity, including how we move, create and – most dramatically – think. So we need to ensure that our education systems are agile enough to respond. If the pace of change feels incomprehensible, it is because it *is* hard to comprehend. From the smug corner of the business lounge, the Davosphere of global pundits, consultants and ex-leaders likes to say we are now in the Fourth Industrial Revolution, which is 'a fusion of overwhelming technological breakthroughs coming almost simultaneously'.

Of course, this is not the first time that technology has changed the jobs we do and the way we live. My ancestors were arrow makers who would have had to dust off their equivalent of CVs when gunpowder came along. The First Industrial Revolution destroyed the job prospects for weavers, and in the Third computers closed human production lines. We are now much more likely to be employed for our brains than our brawn.

But the Fourth Industrial Revolution and its successors will be the first not just to change the type of job we do but our whole relationship with work. The scale and pace of this transformation is already like nothing we have previously experienced: Lyft, Uber, Slack, Deliveroo and Pinterest didn't exist in 2000. Our hyperconnectivity speeds up innovation, and the next phase is going to destroy jobs more quickly than we can create them.

Yet in one respect this next industrial revolution will share a simple and yet overlooked characteristic of the previous three. Through the haze and the tech and the noise, we must understand a simple lesson from history: those who adapt fastest will win. And those who adapt slowest will lose. This is about survival. It is an obvious point. Yet we have failed to grasp it. And as a result we continue to fail to prepare young people to understand it. Perhaps it helps us if we break down the previous industrial revolutions to look at winners and losers.

Britain was the main beneficiary of the First Industrial Revolution, from roughly 1760 to 1840, its economic heft transformed by the steam engine and locomotive. As a result, it rose rapidly up the global league table, developing the political confidence and military might to rule most of the world, for better or – often – worse. Comparable empires that failed to adapt fell rapidly behind, most visibly on the battlefield. Turkey became the 'sick man of Europe'. It will not be the last former empire to carry that unfortunate title.

America became the powerhouse of the twentieth century after the Second Industrial Revolution, from roughly 1870 to

1914, saw mass production and the arrival of electricity, the light bulb, the car and radio. Like the Brits before them, Americans not only invented the things we found we could not live without, they made them. And the rest of us struggled to keep pace with this restless curiosity and patient ambition. Even before the devastation of two world wars, Europe was falling behind. China, which had turned inwards and failed to modernise, found itself humiliated and marginalised.

America was then well placed to be the initial winner of the Third Industrial Revolution, from the 1970s, with the economy automated by the inventions of the personal computer, mobile phone, world wide web and machine intelligence. But emerging superpowers such as India and China are now, in the Fourth Industrial Revolution, finding ways to combine the ingenuity and creativity of the internet with the economic muscle of huge populations and autocratic capitalism. Europe has in contrast seemed clumsy, slow and unwieldy. And America has failed to upgrade its democratic model for the digital age.

As with all technological transformations, there will be winners and losers. 'Technology's impact will feel like a tornado, hitting the rich world first, but eventually sweeping through poorer countries too,' the *Economist* has warned. 'No government is prepared for it.'

So, without waiting for governments, we'll need to find new ways to work and learn. Just as humans evolved in response to the threats and opportunities of the past, we now need to adapt for the Fourth Industrial Revolution, or find ourselves either without work, or working for those who evolved faster than us. We don't know how the class of 2050 will view the world. But if we and they are to thrive, we need a clearer worldview. And that requires us to get better at understanding the future.

Can we discern the outlines of the political and economic landscape that is ahead of us? It might seem foolish to try. But it would surely be more foolish not to try. In analysing geopolitics, diplomats often speak of 'known unknowns' and 'unknown unknowns'. In working out what we need to learn to survive in an age of flux, I think we should anticipate ten known unknowns. All of them underline the urgency of changing what and how we learn. All require new survival skills.

First, the world's population will be larger. Unless fertility rates drop dramatically, we obliterate each other with nuclear or new weapons or future pandemics hit us even harder than Covid-19 did, we will be ten billion by 2050, having only hit one billion in 1800. By 2100, we will take up almost thirty times the space we did in 1500 (when we could all have fitted into one of today's medium-sized cities) and half of us will live in just five countries. India will have almost 1.5 billion people, compared to four hundred million in the US. This growth won't happen everywhere: German, Russian and Japanese populations will shrink. But we will need to be creative about how to sustain this number of humans.

Second, climate change will drive what the *New York Times* has described as 'the greatest wave of global migration the world has seen'. The planet will probably see a bigger temperature increase in the next fifty years than it saw in the last six thousand. By 2070, a fifth of the world will be extremely hot. With every degree of temperature rise, a billion of us will move. As the Sahara spreads south and land fails from Central America to Sudan to the Mekong Delta, hundreds of millions of people will be forced to choose between flight or intense heat, hunger and death. Migration will bring new energy and vigour to ageing societies in the habitable zones. But many countries, facing the overwhelming of political and civic infrastructure by movement on such a scale, will try to seal themselves off. More sensible nations will find new ways to live together. We'll need to learn how to do that.

Third, this combination of climate change and population growth will accelerate the eastward shift of economic and political momentum. China and possibly India are likely to overtake US GDP by 2050. China already has over six times as many STEM graduates as the US. Emerging markets could grow around twice as fast as those of G7 economies. Europe's share of world GDP could fall below 10 per cent by 2050, with the UK down to tenth, France out of the top ten and Italy out of the top twenty, overtaken by faster growing economies like Mexico, Turkey and Vietnam. Those retaining the assumption that the West is the world's centre of gravity will be in for a shock. We will all need to learn fast how to be global.

Fourth, by 2050, we'll also be collectively older. In 1950, the old made up just 11 per cent of the global population. That proportion will overtake the young somewhere around 2075. Advances in healthcare could mean twenty million centenarians by 2100, with pensions – if they still exist – absorbing 10 per cent of global GDP. As we age, we will need to get better at managing our happiness. And we'll need to adapt to working until an older age, and keep learning.

Fifth, power will also move south. Africa will become what the public policy organisation the Brookings Institution calls the 'opportunity for the world'. As it first emerged from the colonial period, it was held back by weak governance and the scars of a time when Europe exported its industrialised weaponry to Africa and imported Africa's resources and talent. But since 2015 the continent has experienced almost thirty changes of political leadership, increasing accountability and democracy. Since 2000, the number of African children in primary school has increased from 60 million to 150 million, with literacy rates up 10 per cent. Connectivity is going to be a great leveller. Studies by the accountancy firm Deloitte have shown that expanding access to four billion people in developing countries will increase productivity by a quarter, GDP by almost three-quarters, and it will create 140 million jobs and lift 160 million people out of poverty. I spent time with farmers in northern Kenya whose livelihoods had been transformed by their ability to get text messages with weather forecasts and market conditions in nearby trading centres. It used to be said that you could give a man a fish and he would not be hungry that day, but give him a fishing rod and teach him to fish and he will never be hungry again. Perhaps the modern equivalent is that we need to give him – or even better, her – a smartphone and an internet connection.

Sixth, there is a risk though that, while the South and East narrow the gap, inequality will increase within our societies: the bottom half of the world's population owns less than 1 per cent of total wealth and the top 1 per cent has just under half. The sad reality is that the Covid-19 pandemic is also likely to hit the more vulnerable hardest. Migrants and refugees already make up the equivalent of the fifth largest country in the world. We'll see greater pressure on food production, especially cereals. By 2050, over half of the world's population will face water scarcity. Without action on the climate crisis, the greatest migration in history will see millions more hungry and angry people on the move. Some gaps will close – including between men and women, and the university-educated and the rest. But the annual World Economic Forum global risk report already lists growing inequality as the biggest risk today. So we'll need the kindness to mind that gap.

Seventh, new weapons make the challenge even more urgent. Our hardwired tendency to apply our restless minds to new ways of killing each other will do to twenty-first century weaponry what it did to the musket, bayonet and pikestaff in previous centuries. Arming new technology such as drones will make warfare easier, cleaner and much more deadly, and reduce our ability to constrain those using it. These weapons will be popular until they are used against us. Elon Musk and other innovators wrote to the UN warning that we are creating risks too fast. There was no meaningful reply. The rules for cyber warfare are dangerously opaque, creating a free-for-all. We have developed strong international systems on conventional, nuclear and chemical weapons, cluster bombs and the arms trade, but we have nothing to manage this new area of rapid growth. As in the past, this risks remaining the case while a few countries have the technological advantage and therefore lack the incentive to agree to restraints. But that will not be the reality for long. These new weapons won't fit into existing international legal systems. They will have to redefine international systems. We'll need to be creative and agile enough to do that, fast.

Eighth, international systems for managing these challenges – infectious disease, migration, the health of the global economy, climate change, terrorism, new weapons – will fail before they start to succeed. We will learn the hard way that these threats cannot be tackled within national borders alone. Economic downturns cause countries to look inward at the very moment when it is most important to look outward. The era of US-led global cooperation began after the Second World War with the construction of the international architecture we retain today. It peaked in 1989 with the fall of the Berlin Wall, and it ended in 2016 with the election of Donald Trump. The postwar international system was built with immense sacrifice by previous generations to protect us from the dangerous individual or ideology that believes it alone has the answers. The international bodies that were created are full of hard-working professionals, but they have been tested in the twenty-first century by American retreat, Russian disruption and the east-ward shift of power. Now Covid-19 has become another weapon in the armoury of those who believe that global challenges require national solutions. A pandemic that exposed the weakness of systems for international cooperation led many politicians to advocate national distancing alongside the social kind. Fixing the global scaffolding will take serious determination and patience. We will have to ensure that we enfranchise more of the world's population. Reform is especially important in a time of great power rivalry, but that's when the great powers least want it. In their absence, who drives the process, convenes the meetings, decides who gets in the room? And can those in the room claim any more to speak for those outside? Events will get in the way. As the former US president Dwight Eisenhower warned, long before social media and Oval Office tweeting proved him right, 'What is important is seldom urgent and what is urgent is seldom important.'

In 2017 I wrote a report for the United Nations secretary-general that set out the steps the international community needed to take – urgently – to prepare itself for massive geopolitical and social transformation as a result of such rapid technological and social change. Evidently the United Nations could not take those steps. And as it staggered past its seventy-fifth birthday in 2020, the UN has more than underlying health conditions: it was orphaned and abused by the Trump White House and found itself paralysed by the great pandemic that began in 2020.

Ninth, national systems will also struggle to cope. As Bill Gates has said, we overestimate the short-term implications of technology, but we underestimate the long-term impact. We have seen how most governments – whether democratic or authoritarian – struggled with Covid-19. The economic and reputational hit will be greater in some countries than others, moving them up or down the league tables. And with so much energy absorbed by the immediate crisis response, there was less to contain viruses such as extremism, radicalisation and injustice, or to protect refugees and civilians under attack. Most governments are finding it harder and harder to be strategic. Before I worked in 10 Downing Street, I always believed that there would be a room of brilliant thinkers in the basement, with a plan. There wasn't.

As former prime minister Gordon Brown told me in his Kirkcaldy study, surrounded by mountains of books, 'I think what happened during the pandemic was that leaders wanted to do something. But they underestimated the scale of the challenge. They got themselves so locked into national issues that they couldn't understand that this is a global pandemic that cannot be solved without coordinated action. As a leader, you need to be able to step back and ask "What am I trying to achieve? And how can we achieve what I'm trying to achieve? And are other people trying to achieve it with me?" You have got to find a way of talking about the big picture and the big picture matters. But you also need a mastery of detail. That's getting harder. We used to say that the past was a foreign country. For policymakers and leaders today, the present is a foreign country.' Perhaps one of the few things that Presidents Obama and Trump could agree on is that it is getting harder and harder to govern. This is due to a combination of the speed of events, relentless scrutiny of decisions, distrust in authority and failures of governance. Without a new social contract between citizens and those who govern on their behalf, we won't be ready for the greater challenges ahead. So the future is also a foreign country.

Meanwhile, technology companies will continue to grow in size and power. In 2021, Twitter effectively switched off a US president, and one who had become one of their most-followed users. Already, Apple is a bigger economy than Nigeria and Microsoft is bigger than Egypt. Automation and artificial intelligence will affect countries in different ways and at different paces. In some sectors and regions, there will be great economic benefits to users and businesses. In others they will be destructive. We'll need the creativity and ingenuity to compete with the machines.

A final trend. The sharing economy, through which people use the internet and mobile devices to access goods and services, is here to stay: there were gasps of surprise when PricewaterhouseCoopers estimated that it would grow twentyfold between 2015 and 2025, but this now looks conservative. As entrepreneur Tom Goodwin has written, 'Uber, the world's largest taxi company, owns no vehicles. Facebook, the world's most popular media owner, creates no content. Alibaba, the most valuable retailer, has no inventory. And Airbnb, the world's largest accommodation provider, owns no real estate.' So we'll need to retain the agility to find our place in a more fluid, less material economy. Our curiosity will become even more essential to our ability to cope. More importantly, it will be crucial to our ability to shape the potential of the sharing economy in a way that strengthens society.

Population growth. Climate crisis. Demographic earthquakes. Geopolitical power shifts. New weapons. Failure of international cooperation and national strategies. The onward march of Big Tech and authoritarian capitalism. The inequality that will accompany the transition to a sharing economy. A frightening combination.

Facing these megatrends, it should be no surprise that most of us feel unprepared and apprehensive, that we worry that the world is somehow out of control. It is.

So what do we *really* need to know to survive?

PART ONE

Ten Survival Skills

1

How to Take Back Control

On what principle is it that when we see nothing but
improvement behind us, we are to expect nothing
but deterioration before us?

Thomas Macaulay

In the midst of all this change and flux, we could be forgiven
for feeling overwhelmed. Whether you agree with them or
not, political campaigns such as Brexit and Donald Trump's
'Make America Great Again' have found ways to play effectively to this sense of dislocation and the desire to find a more
stable footing.

Without leaving any more trading blocs or electing any
more Trumps, how do we find better ways to feel more in
control, or at least less out of control? How can we get better
at preparing ourselves more effectively for whatever the future
holds? We can do it by getting better at anticipating and
understanding what lies ahead; by developing the resilience
we will need for crises and setbacks; and by taking back

control of our time and our learning, so that we have the survival skills we will need.

Anticipating the future sounds daunting. The good news is that we have a head start. Predicting change and evolving in preparation for it is something we do better than the competition. We can't outrun a cheetah or out-wrestle a silverback. But *Homo sapiens* have always been better planners than any other species.

As hunter-gatherers, we learned to predict where and when animals could be caught and trees would give most fruit. By the time of the Romans, those practised in the art of augury were basing key military decisions on their interpretations of the world of birds. Over the course of thousands of years of recorded history, and thanks to brave minds such as that of the mathematician and astronomer Nicolaus Copernicus, human beings became experts at predicting the motions of the stars and planets. Leaps forward such as calculus and modern statistics laid the basis for modern data science, machine learning and predictive analysis, covering everything from where to graze cows to divining weather patterns and credit risks, as well as web searches or sporting line-ups. That market in predictive analysis will reach up to $8.5 billion by 2025.

But knowing the future is not brain science. Philip Tetlock, the co-author of *Superforecasting*, has sought to demonstrate that people can become better at forecasting and they can teach others how to forecast more effectively. If he is right, why are we not spending as much time and effort on this skill as we do in the gym?

Dubai's futurist-in-chief, Noah Raford, has been described as the 'vizier and provocateur' of the future. Walking through the spectacular Museum of the Future in Dubai, he told me we all have a psychological need to feel we can understand and somehow control the future, but we must understand that the world is stranger than we conceive. It is irresolvably uncertain and complex. For Raford, the most important aspect of his work is not prediction but 'to use the lens of the future to help us perceive the present more clearly'. A better understanding of what is likely to happen will allow us to make the reforms we need now.

'Can the rest of us really become forecasters ourselves?' I asked, feeling overawed by the robot that had just greeted me by name, remembering me from an earlier visit.

Noah grinned as he high-fived 'Poppy'. 'Absolutely. This is not about gadgets and tech. Forecasters aren't trying to predict sports results, but how people will react to the world around them. The key is to proceed with intention.' Once you take the conscious step to examine and try to anticipate trends, you are already better prepared.

How do we get better at anticipating what lies ahead in order to prepare ourselves for it?

We need to start with humility and caution. For all our successes in planning and programming, humans, including most soothsayers and astrologists, have just as often proved poor at predicting the future. Futurist Ray Kurzweil has been one of the more successful. But while getting it right on the threat to privacy from tech and data, and with health gadgets such as fitness watches, he predicted in 1999 that human life expectancy would rise to 'over a hundred' by 2019, and that books would be dead. Whether you are reading this on paper or onscreen, both are evidently not true. Predictably, those who get forecasts right normally get plenty wrong too.

Predictions of global population passing eight billion, flying cars, electronic voting, Chinese democracy, humans on Mars, replacement of fossil fuels, three-day weeks, smart homes, anti-gravity belts and China overtaking America economically are frequently revised. In 1968, the MIT political science professor Ithiel de Sola Pool predicted that better communication, easier translation and greater understanding of human motivations would mean that 'by the year 2018 nationalism should be a waning force in the world'. If only. The inventor Nikola Tesla, a visionary on so much, predicted that 'twenty-first-century humans would know better than to pollute their bodies with harmful substances like caffeine and nicotine'. Warren Buffett is right to warn us to always consider the motives of those making the prediction. 'Don't ask the barber whether you need a haircut.'

The discouraging point about these examples is that it is impossible to get it right all the time. The encouraging point is that there is no reason why we can't give it a shot ourselves.

To get better at doing so, we can get better at understanding and monitoring the megatrends I described in the introduction. Like the physicist John Dalton, born in 1766, who relaxed by using homemade equipment to record and predict the weather. His fifty-seven years of records were the earliest such dataset in the UK, forming part of his studies of humidity, temperature, atmospheric pressure and wind. From his data, the science of meteorology was founded. But we will also sometimes need, in journalist Nik Gowing's words, to let ourselves 'think the unthinkable'. Like theoretical physicist Peter Higgs and his team who in 1964 predicted the existence of the Higgs boson elementary particle through the Higgs mechanism. Yet it wasn't until 2012 that the particle was confirmed to exist and Higgs won the Nobel Prize. We can help ourselves to think the unthinkable by seeking out ideas we disagree with and consciously challenging our existing assumptions about the world.

We can also get better at making connections between the past and the future. Futurist Anne Lise Kjaer says that, like archaeologists, futurists 'use artefacts from the present and try to connect the dots into interesting narratives in the future. When it comes to the future, you have two choices. You can sit back and think, It's not happening to me, and build a great big wall to keep out all the bad news. Or you can build windmills and harness the winds of change. It's a huge opportunity to educate the parents of the next generation, not just the children.'

A key part of taking control, then, is also to anticipate and prepare for setbacks and crises. Our freshly polished antennae will tell us to assume one thing: surprises and uncertainty are inevitable. And that no plan survives contact with the enemy.

Traditionally it was believed that the ability to come through adversity could only be observed or acquired in times of stress. To develop the capacity to bounce back, you needed something to bounce back from, whether a violent threat, trauma or difficult childhood or home life. But recent work by psychologists such as Norman Garmezy and Emmy Werner has shown that resilience can also be learned. Sometimes that happens because of the good fortune of contact with a supportive teacher or mentor. Yet often it is about how we perceive our sense of agency, whether we think that we rather than our circumstances affect our way of being in the world. People who can meet the world on their own terms are able to become more autonomous and independent. Like in the Tennyson poem, more resilient people think of themselves as masters of their own fate.

George Bonanno, a clinical psychologist at Columbia University, has been studying resilience for twenty-five years, trying to figure out why some people are better able to handle stress. He told the *New Yorker* that a key factor is whether an individual is able to understand an event as traumatic or as an opportunity to learn and grow. Every frightening event can be traumatic or not to the person experiencing it. What matters is how we explain it to ourselves.

Can we train ourselves to be better prepared for shocks, and to process more effectively those we have already experienced? Yes. The first step is to recognise our ability to regulate our emotions more successfully. The next, University of Pennsylvania positive psychologist Martin Seligman told me, is to reframe the way we describe negative events to ourselves. Bad events aren't our fault or indicative of some wider failure in us. We have agency to change our situation. If we view adversity as a challenge to learn from, we are more likely to come through it. Resilience is our ability to bounce back from stress, to adapt to challenging circumstances, to thrive in adversity. It requires self-awareness and awareness of others. People who are more resilient will tend to see setbacks as temporary, specific to a moment or situation, and not all about them. They are less likely to see setbacks as permanent, more general, or very much about them. They will feel more in control. There is an ongoing competition between factors that increase our grit and those that increase our stress. So we have to keep working at our preparedness. From my experience, for example as ambassador in Beirut and Syria and working as an adviser to the G20 after the 2008 financial crash, there are ways we can train ourselves to do this.

First, practise thinking carefully under pressure. This was the mantra of England's World Cup-winning rugby union coach Clive Woodward. It sounds simple, but we know how quickly that clarity goes when a crisis hits. People act very differently under pressure. Our adrenalin picks up, we take hasty decisions, we panic. As with a rugby team, it is possible to practise thinking carefully under pressure, in the way that Woodward's 2003 team did in the decisive closing moments of the final. To prepare for the big crises, we can get better at not ruminating over the small ones, recognising what is beyond our control, and focusing on the areas where we do actually have agency and influence, including in managing our emotions. If we want to take back control, that's a good place to start.

Second, during a crisis you need more than ever to know where the scaffolding is. Who are the people or what are the habits that you will most rely on when it gets tough? We can prepare ourselves by investing more heavily in those load-bearing relationships and practices in advance. Look after the basics.

Third, communication is even more important during times of stress. We need to make even more of an effort to listen. 'How do we fix this together?' is a good crisis question. Make people part of the solution. Don't assume that because you've said it once, everyone has heard and understood it. Ask the right questions, of yourself and those around you. Who here disagrees with this approach? If no one, that's a problem.

The next step to taking back control is to regain more control of our time.

Time is a powerful currency. We all know that the time we have is finite. As we get older, we feel the sand falling through the glass. A good back-of-the-envelope exercise is to think about what you would really love to be doing today, or this week. How much time would you set aside, in an ideal world, for doing the things and spending time with the people that bring you joy? And then write in the column alongside it a breakdown of what you are *really* doing today, or this week.

The difference between the two can be stark and discouraging. But it can at least help us to see where we can make changes.

Sometimes, it may just be that we need to slow down. My successor as the prime minister's foreign policy adviser in 10 Downing Street was my friend John Casson. He is the smartest foreign policy thinker I've met. After a brilliant stint there he went to Cairo as ambassador. During a tough political period he did an extraordinary job, amassing over a million Twitter followers along the way.

And then he stepped back, devoting his time to what he describes as stillness, attention, self-awareness, rest and letting go. John told me that he had learned in this time that 'we become what we pay attention to. Ruthlessly eliminate hurry. Pursue stillness. Pay deep attention.' He introduced me to the blessing of John O'Donohue:

This is the time to be slow
Lie low to the wall
Until the bitter weather passes

Try, as best you can, not to let
The wire brush of doubt
Scrape from your heart
All sense of yourself
And your hesitant light.

If you remain generous,
Time will come good;
And you will find your feet
Again on fresh pastures of promise,
Where the air will be kind
And blushed with beginning.

Taking back control of more of our time might also mean reducing the time we choose to sell to our employer. How can we increase our financial independence and buy ourselves more time? What do we really need to live on each year? What are the savings we can make now to get there? There are shelves of books with the answers to those challenges. They boil down to two unexciting but decent ideas: reduce your debt and own less.

'Every time you borrow money, you're robbing your future self,' counsels personal finance expert Nathan W. Morris. This involves tough choices. We have been told relentlessly that the key to happiness is to spend. Yet we know it isn't. 'Too many people spend money they haven't earned to buy things they don't want to impress people they don't like,' as Will Rogers put it. Shopping itself stimulates our primal hunter-gatherer urges, making us worse judges of what we actually need: look around your home or the next boot fair for all those unwanted impulse buys.

The science suggests that, if you know what to do with it, a certain amount of money helps, in that it can reduce unhappiness. But we can't purchase real happiness, because we're never satisfied. We overestimate the pleasure having more brings, a condition that economists call the 'hedonic treadmill'.

Sometimes the extra money brings new stress: a longer commute or a new set of richer neighbours to envy. Humans compete: this is a survival skill that has helped us through much of history. But once we have met our basic needs, it may hold us back. In *Stumbling on Happiness*, Harvard professor Daniel Gilbert says, 'Once you have your human needs met, a lot more money doesn't make for a lot more happiness.' When we think of life in terms of time rather than cash, we might find that what we most value is a weekly walk in nature, a decent internet connection, a regular lunch with friends and the occasional break exploring somewhere new. 'Wealth consists not in having great possessions, but in having few wants,' counselled the Greek Stoic philosopher and former slave Epictetus.

How we perceive our relative wealth has more impact on our happiness than the wealth itself. We can at least stop comparing ourselves to the Clooneys, Jeff Bezos or the latest winner of the lottery, and start comparing ourselves to what the Dartmouth College economist Erzo Luttmer has called 'similar others' – the people we work or grew up with, old friends and classmates. Sonja Lyubomirsky, a psychology professor at the University of California, argues that happy people aren't bothered by the successes of others. When she asked less happy people who they compared themselves with, 'they went on and on'. But 'the happy people didn't know what we were talking about'. They dared not to compare.

We can buy more time by focusing on what we need rather than what we want. People overestimate the pleasure they'll get from things, and underestimate the pleasure they'll get from experiences. The science suggests that sometimes it can often be the things that don't last that create the most lasting happiness. Students rated experiences (a night out or holiday) more highly than possessions. One reason may be that experiences tend to get better, not diminish, as you recall them. People who are happiest are often those who are able to think of everything as an experience, including spending.

The most Googled financial questions are how to save, invest and retire. The answer to those searches may be simpler than we think, and combines all three. Peter Adeney, a Canadian blogger and software engineer, is better known as Mr. Money Mustache, having retired at the age of thirty by minimising his spending and focusing relentlessly on the goal of financial independence. Describing the typical middle-class lifestyle as 'an exploding volcano of wastefulness', he has argued that by spending less money and owning fewer physical possessions we can give ourselves a better shot at happiness.

At the heart of his approach is the idea that you calculate what you really need to live on, and reduce your spending to that. His followers are ruthless in eliminating debt, tracking their finances and setting aside an emergency fund to free them from anxiety about the what-ifs of life. The goal of retirement is not to retire from doing things. Quite the opposite. It frees them to do the work they love, to find purpose. 'You can only become truly accomplished at something you love,' said Maya Angelou. 'Don't make money your goal. Instead, pursue the things you love doing, and then do them so well that people can't take their eyes off you.'

She is right. If we think of time rather than money as the key to the life we aspire to, we have more chance of being liberated to live that life.

Taking back control also requires us to take greater responsibility for the skills we have, to ensure that we are as well prepared as possible for what lies ahead. 'Live as if you were to die tomorrow. Learn as though you would live forever,' advised Mahatma Gandhi. And yet most of us blunder through life without taking the time and energy to develop ourselves. We act as though our education ends when we leave formal education.

We no longer have that luxury.

We are faced with a workplace that in just twenty years' time will be unrecognisable, thanks to digitalisation and automation transforming every industry. Already, business leaders are saying that the current education model is not providing young people with the social and emotional skills that are required. They need more people who can solve unstructured problems, work with new information, share and critically evaluate new knowledge. Working well with a team is more important to them than previously.

There is a story that the painter Titian put down his brush at the age of eighty-six saying, 'I think I am beginning to learn how to paint.' Learning should not stop at the school gate: we all need to learn that we still have a lot to learn. Social entrepreneur and educator Tony Bury told me that the students who made the most significant contributions to society were those who gained 'learning potential'. Individuals need to be able to 'embark on a learning curve and disembark onto the next learning curve'.

Oxford experts found that adult education is a 'permanent national necessity'. Yet there is massive variation from country to country in the provision of lifelong learning, and in adults' use of literacy, numeracy and problem solving. Few come close to the Danes, a third of whom take part in the Danish government's lifelong learning scheme, with two weeks of skills training a year for all citizens. Giving a speech at Copenhagen University, I was perplexed to find that most of the students on the scheme were pensioners.

Applying yourself to something new or hard also makes you happier. We're addicted to challenges that help us develop what psychologist Mihaly Csikszentmihalyi called a state of 'flow': total absorption in something that stretches you to the mental or physical limit. One study by University of British Columbia researchers has suggested that workers would be happy to give up as much as a 20 per cent pay rise if it meant learning more new things and having more variety.

So we need to cultivate the mindset that we are learning all the time, too. As Lee Kuan Yew, Singapore's then prime minister, put it: 'I have never ceased to be a student. I have never ceased to learn.' The island's 'SkillsFuture' programme has proved a brilliant example of this. Every Singaporean receives the equivalent of $300 to 'pursue lifelong learning, build personal mastery and pursue their passion'. There is an online databank of ten thousand courses. The designer of the programme described it to me not as education but as 'a national movement to provide Singaporeans with the opportunities to develop their fullest potential throughout life, regardless of their starting points'. As well as its databank of courses, SkillsFuture has a huge hub of resources for skills training, knowledge development and career planning. The minister who oversaw its development told me that it was conceived in response to a realisation that Singapore's education system – very successful at driving up exam results – had widened the social divide and was no longer encouraging social mobility. A highly skilled workforce was not enough. 'We upgraded education, accessible to all, to create a more equitable society and to ensure a wider pool of talent.'

The Singaporean government realised that a generation of people had been trained to pass exams. They were smart enough to see that people needed further support to develop character, citizenship and wellbeing. As part of the effort, the country's media were encouraged to look beyond academic league tables. Many other countries would benefit from such an enlightened approach.

The case for lifelong learning is not just about the demands of the economy and having a skilled workforce. Or about making ourselves more employable. Our NYU research also showed that we need to keep learning to improve our wellbeing and self-confidence; and to enhance our ability to interact with new people and new ideas. It feeds the education of head, hand and heart that this book advocates. Neuroscientists also believe that the best way to retain plasticity in our brains is to keep learning. Paying real attention to developing a new skill is the closest our brains get to a workout.

Lifelong learning like this is easier when we recognise the way that our intelligence changes over a lifetime. Writing in the 1940s, British psychologist Raymond Cattell highlighted the difference between fluid and crystallised intelligence. We tend to have more fluid intelligence when we are younger: it is the speed, creativity and agility to solve new problems. That diminishes as we age, to be replaced by crystallised intelligence. This is about using the knowledge we have gained, joining the dots between themes and ideas, and distilling it into wisdom. Are you in a career that relies more on fluid intelligence or crystallised intelligence? Which did your education support? What have you learned this year that developed your crystallised intelligence?

With the prospect of working in several fields over a lifetime, how therefore do we ensure that we have the right mindset and toolkit to keep adapting and learning? There is no standard approach. This is not a factory line. We all learn differently.

You may be a visual learner, needing images, mind maps and graphs to take in information: in that case, you might now go to YouTube or Netflix to pick up new skills. Kenyan javelin thrower Julius Yego came eighth in the 2012 London Olympics despite never having had a coach or a lesson – Kenya focuses on its runners. He had taught himself via the YouTube videos of Norwegian javelin thrower Andreas Thorkildsen. Yego became world champion in 2015 and has the third longest recorded javelin throw.

You may be more auditory, understanding most when something is spoken or talked through: you may thrive in seminars and lectures. Before the printing press, this would have been a significant part of how universities passed on knowledge. Students would listen while the text was read out. In some universities it is still the basic model.

You may be a learner who takes in most of your information by reading and writing. Since the printing press, this has been the main way that universities have passed on knowledge. The quality of the library became more important than the quality of the lecture.

Or you may be more of a kinaesthetic learner, taking in new ideas by moving, creating and practising: you might work best to music or by switching environments.

Most of us are a mixture of all four.

You might be more left-brained, comfortable with logical, mathematical and analytical approaches, and good on the detail. Or right-brained, tending towards the creative, artistic, intuitive and the big picture. Our NYU research dug further into what this means for how we acquire and retain new information and skills. If you are strongly *left-brained*, you will enjoy learning more if you have a clear step-by-step syllabus with targets along the way. You will like the structure of a well-organised daily timetable, regular revision sessions and clear explanations. You are more likely to learn through real-life tasks that require discussion and teamwork. Sound familiar? Left-brained learners were more likely to adapt well to online learning that provided the scaffolding to structure their time during the pandemic lockdown, and detailed content.

If you are strongly *right-brained*, you will be more likely to enjoy learning by working with pictures, videos, graphs, mind maps and colour coding. You might enjoy role-plays and drama as a means of retaining new ideas. You will be less likely to want or need much structure – perhaps an overview or a set of learning goals. This type of learner may prefer learning on the job to on a training course. You're less interested in the theory and want to get on and try it out.

Right-brained learners adapted well to online learning during lockdown when they had the freedom to pursue ideas and structure their own curiosity. They were more likely to feel liberated from the school timetable.

Again, many of us are a mixture. You probably recognise parts of yourself in both descriptions, or see ways that you move between the two. Much of the best online learning from schools during the pandemic found ways to combine the structure and the freedom. It worked harder to identify the essential content – information that had to be understood and retained. But then to give students different pathways to retaining it: drawing a poster, putting it to music, making a video. In the massively expanding online adult education market, the focus has been more left-brained. It is easier to master structured content from a YouTube video than to interact with others to help you explore and understand it. But, again, the most effective content is increasingly blending the two approaches, introducing more visuals, colour, music.

Another way to consider learning styles is to reflect on the balance between activists, reflectors, theorists and pragmatists. Activists learn best when they are enthused. They want to be stimulated by new experiences. 'Let's give it a go.' You can picture the activist at the training day, grabbing the Post-it note and heading to the whiteboard.

Reflectors are more likely to stand back. They want data, and time to assess, challenge, revise. 'Let's not rush ahead of ourselves.' Most academics tend to be reflectors.

Theorists like to think through problems in a logical way. They are better able to detach themselves from the emotional aspects of a challenge. 'Does it make sense?'

Finally, pragmatists like to seek out new ways of thinking, but are impatient to get on with practical application of what they have learned. 'If it works, great.'

Of the leaders I observed close up, Nicolas Sarkozy was an activist, Angela Merkel a reflector, Barack Obama a theorist and Gordon Brown a pragmatist.

Just recognising these differences helps us to take back control of our lifelong education. Think of people you have studied or worked with, and the way that they naturally fit into the different categories. And then think of how frustrating it is when you are expected to learn in a way that doesn't fit for you. There are generations of young people whose learning was inhibited by being forced into classrooms, lecture halls and curricula that were too hierarchical, or too left-brained.

Looking at a broad range of global students, our NYU research concluded that the most effective approach is that of allowing students – that is, all of us – more autonomy and flexibility over how we best learn and engage. We gave students a task with different options for solving it: working in a team; researching the detail and breaking it down; using images or video to explain it. The best outcomes came from giving students this freedom but then bringing them together again to share what they had learned.

There are plenty more practical ways to upgrade our learning capacity. Mark Fletcher has spent a lifetime teaching and writing about brain-friendly learning. As his son, I'm one of many beneficiaries of his methods. While we all learn in different ways, he offers a menu to take our learning to a different level. He says, 'It has to start with taking responsibility. You are the only person who can learn for you. Keep a learning diary, taking time to set out what you want to learn, when and how, and checking back on progress. Take regular breaks – forty-five minutes is all most of us can cope with in one session. And build in regular sessions to review and process what you have learned.' If we leave learning to chance, it won't happen. We don't learn much when we are bored or anxious.

For learners who are over twenty-one years old, the approach to learning that our NYU research found was most consistently effective, and which this book seeks to offer, is to observe, learn, practise, teach. For those facing the prospect of adapting their skills to new roles, or learning completely new skills, this offers a simpler and less daunting pathway.

We start with observation, for instance of an interesting example of a survival skill or the research that explains it. You might decide that you need to improve your ability to sift through increasing amounts of online information, or to get better at thinking critically about the news that is presented to you. Look for role models. Who does this well? What is their method? One of the most powerful questions to ask young people – or indeed any of us still working things out – when they describe an experience is 'What did you learn?' It often prompts the most thoughtful re-processing, and helps people through those stages of observation.

Then we try to distil that into a memorable way of retaining it. To take the observations on learning resilience from earlier in this chapter, you could break them down to the need for more focus on preparation, practice and people. Then get out and actually practise. Consciously put yourself in situations that will require you to use this skill. Set yourself manageable tests and goals. Check your progress. Ask others to hold you to account for it. And to really learn the skill, teach it to someone.

This is how the great apprentices learned, from painters to plasterers to mathematicians to hairdressers. This book will apply that framework to the new – and not so new – survival skills.

Like Titian, the American civil rights activist Maya Angelou saw education as a journey not a destination. She overcame poverty and abuse to become a brilliant author and poet. In a letter to her daughter, she talked about what she had discovered on the way. 'I've learned that no matter what happens, or how bad it seems today, life does go on, and it will be better tomorrow. I've learned that you can tell a lot about a person by the way he/she handles these three things: a rainy day, lost luggage, and tangled Christmas tree lights ... I've learned that whenever I decide something with an open heart, I usually make the right decision. I've learned that even when I have pains, I don't have to be one. I've learned that every day you should reach out and touch someone ... I've learned that I still have a lot to learn.'

The good news is that the key to this lifelong learning may be more attainable than we think. What do I need to know? What can I learn that will set me apart in future? What are my least developed survival skills?

Ask ourselves these questions and we are able to shape our own curriculum. We can turn ourselves into constant learners, consciously evolving. We can think practically about our learning: keeping a learning diary, reflecting on our learning style, working out the skills so that we can move through those gears from observer to teacher. And we can wonder what might be our equivalent of Titian's brush? What at the end of our lives will we be keenest to still be learning? We develop the survival skill of curiosity.

2

How to Be Curious

Be less curious about people and
more curious about ideas.

Marie Curie

At the age of only thirty-two, Melanie Perkins is Australia's youngest billionaire, and the third richest woman in the country. The daughter of a Malaysian engineer of Filipino and Sri Lankan descent, she was the force behind Canva, a graphic design startup that is now worth almost $9 billion, having doubled in value during the Covid-19 lockdown. Melanie estimates that it was rejected by over a hundred investors, until she took up kite surfing in order to network with venture capitalists. 'It was like, risk: serious damage; reward: start company,' Perkins told *Forbes* magazine. 'If you get your foot in the door just a tiny bit, you have to kind of wedge it all the way in.'

Canva has liberated huge amounts of ingenuity and creativity by allowing anyone with an idea (maybe for a business, a presentation, or even just a party invitation) to design their own graphics in support. Previously the software would have been too expensive and hard to master. Now anyone can become a graphic designer. The over twenty million people who have used the service seem to agree, but the money does not appear to have gone to Melanie's head: her co-founder proposed to her in 2019 with a $30 engagement ring, while they were backpacking in Turkey.

Melanie says she learned ingenuity as a teenager after she started her first business at the age of fourteen, making scarves. Perhaps her early aspirations to be a professional ice skater – and the early mornings and patient practice – helped. But there still needed to be a moment of genius. She came up with the idea for Canva as a university student, while daydreaming on her mother's couch in Perth. How often do we hear successful entrepreneurs and inventors talk about such a moment, when the light bulb switches on and their destiny changes?

As we've seen, what we do as humans, and how we do it, is changing at a faster pace than at any time in history. Our survival to date has been based on our success at managing these moments of transformation. It requires ceaseless creativity and innovation. The Stone Age didn't end because we ran out of stones, but because we found a better way to make what we needed. Our curiosity is what has enabled us to exert greater control over the world around us. It is an essential survival skill.

Humans are curious by nature. A 2016 American study demonstrated that curiosity could lead humans to expose themselves to aversive stimuli (even electric shocks) for no apparent benefits. We possess an inherent desire to resolve uncertainty: research has shown that when facing something uncertain and feeling curious, we can act to resolve the uncertainty even if we expect negative consequences. So while at times we need to manage our constant desire to solve riddles, we are hardwired to tinker and to question.

As chief executive of the British Library, Roly Keating gets to see this every day. 'From about 9 o'clock onwards this queue of people begins to form,' he says. 'And you don't know what they're there to do, but they've all got a mission of some kind to find out. They're all ages, all backgrounds. And it stretches right across the piazza almost back to the road. And that is like a daily reassurance that somewhere in those vaults, somewhere in that extraordinary building, is the answer to everyone's question, or at least the thing they will discover that will prompt the next question. Every morning, lorries arrive here like it is a farm or factory, with everything published that day. That's liberating because it's not my judgement or my curators' judgement as to what to preserve, it's just a commitment to preserve and then let the curious-minded come in to make their choices.'

I put it to Roly that in the Middle East they still tell stories of the River Tigris flowing black with the ink of the books from the Bayt al-Hikmah (or House of Wisdom) that had been destroyed by Hulagu Khan's Mongol invaders in 1258. 'What would you run back into the library to save from a fire?' I asked. He looked pained. 'It's a bit distressing. I'd be reaching out for the Magna Carta in one direction and then I'd see Jane Austen's manuscripts over there, and then, oh my god, it's the Mozart autograph manuscript which we really ought to try and look after. And then I'll notice Sultan Babar's Koran over there, and I guess we mustn't forget the earliest printed book in the collection from 862 from China. Or *Abbey Road*. It is almost painfully hard to be selective there. Because the truth is that the real journey of knowledge is about fostering the skills of the last people, whatever survives, as it were, this fire.

To make sure they've at least got the curiosity, knowing how to ask the right questions, how to piece together the evidence of what is lost.'

The twenty-first century is a great time to be curious. 'If you are hungry for food, you are prepared to hunt high and low for it. If you are hungry for information, it is the same,' says Stephen Fry. 'Information is all around us, now more than ever before in human history. You barely have to stir or incommode yourself to find things out. The only reason people do not know much is because they do not care to know. They are incurious. Incuriosity is the oddest and most foolish failing there is.'

The greatest educators have always sought to inspire curiosity. And the greatest innovators have continually harnessed it. When I asked him to identify the most important survival skills, the entrepreneur Richard Branson told me that insatiable curiosity came top of the list. We must remain curious in a time of too much certainty. Why?

First, as we have seen, curiosity helps us to learn. Research by the University of California in 2014 looked at what exactly goes on in the brain when our curiosity is aroused. They rated how curious participants were to learn the answers to over a hundred trivia questions, while using fMRI scans to examine what their brains were doing. The researchers found that curiosity prepares the brain for learning and makes that learning more rewarding. In *Curious: The Desire to Know and Why Your Future Depends On It*, Ian Leslie has drawn on research from psychology, economics, education and business to show that curiosity is a mental muscle. At its best it combines intelligence, persistence and hunger for novelty. Like other muscles, it atrophies without exercise.

Second, curiosity helps us to innovate and to succeed professionally. The most successful companies understand this need for people who can explore, tinker and create. It is not about classic IQ: many leading companies are now using brain-teasers and riddles to recruit. They are less concerned with whether we can answer a question, but how we approach a complex situation. In 2004 an anonymous billboard appeared in Silicon Valley, posing a maths puzzle. Those tempted to tackle it found further equations online, which led them to a request to send their CV to Google. Eric Schmidt, Google's then CEO, told me in 2015 that 'unlike most companies, we are not looking for know-it-alls with perfect academic records. This company only succeeds if it runs on questions, not answers.'

When we are curious about our work or study, we achieve more. In her advice to employees on how to succeed, PepsiCo CEO Indra Nooyi urged them to 'remain a lifelong student. Don't lose that curiosity.' This is a big change from a business model that has rewarded people for getting on with it rather than asking awkward questions. Harvard Business School now has a whole module on how to hire for curiosity.

Third, curiosity makes us happier and healthier. This makes it easier for us to tolerate anxiety and uncertainty. It helps us choose the healthier lifestyle we need. In one study, simply by posting trivia questions near a university building's elevators and posting the answers in the stairwell, researchers created a 10 per cent increase in the use of stairs. In another, 'they increased the purchase of fresh produce by placing a joke on the placard describing the fruit or vegetable and printing the punchline on the bag's closures'. Curiosity also strengthens our relationships, which are central to our wellbeing. We see people as warmer and more attractive when they show genuine interest in us.

Edith Wharton – American novelist and the first woman to win a Pulitzer – agreed. 'In spite of illness, in spite even of the archenemy sorrow, one can remain alive long past the usual date of disintegration if one is unafraid of change, insatiable in intellectual curiosity, interested in big things, and happy in small ways.'

Finally, curiosity helps us coexist: it makes it easier to do what we will come to in Chapter 6: live together. My Emirati friend, diplomat and polymath Omar Ghobash, wrote a brilliant book called *Letters to a Young Muslim*. In it he pens a series of honest and thoughtful letters to his son Saif about the search for the voice of his own father – a government minister who was assassinated when Omar was just six. In the process, he reflects on the complex modern debates about identity, belief and nation. If we are to resolve them, he argues, it is vital to create and protect the space to debate and compromise. Omar warns his son in words that should resonate across the Middle East and beyond: 'I want you to be on the lookout for people who tell you with unerring conviction what you should do and think.'

Globally, we seem to be in a period when there is too much certainty and too little curiosity. People are finding themselves drawn further into echo chambers, in which they hear only the views of those with whom they already agree. As anyone with a smartphone or a tablet knows, they can unlock extraordinary and exciting potential. But they can also make us idler and more apathetic or distracted. It is all too easy not to care, to see it all as too difficult, to swallow the easy conspiracy, or simply to oppose. The internet has given a voice to the angry and intolerant. It has become harder to find those ready to fight for something, as opposed to against something. It is often easier to destroy than it is to build.

But curiosity expands the empathy that we need to reach across those lines. It helps us understand the different experiences of others. It is the key to unlocking wider emotional intelligence: our ability to tune into emotions, understand them and respond to them.

I have spent too much time online trying to trace the way that young people are radicalised. The language is similar, whether in Kansas or Kabul. 'We can take you back to a time when you were more powerful, more respected, greater. Life was better before these new ideas and people arrived in your community.' It is getting harder to hold the territory that Isis and its future emulators attack as the 'grey zone', or the far right and far left attack as the liberal consensus. But this space is where curiosity breathes. It is the space where individual freedom to think and speak is cherished and diverse communities interact. It is by its nature imperfect, a work in progress. So we must also be ready to fight harder to uphold the curiosity we share than the intolerant – in all our societies – are fighting to destroy it.

I asked Instagram influencer Jeremy Jauncey why he was such an advocate of travel. I expected an Instagram-friendly answer about finding oneself. Instead he argued passionately that travel is a universal language: 'It is the thing that can help educate people that the racism and bigotry and inequality and hatred and all these prejudices can be completely dissolved. You just have to get out and spend time in a person's country and walk a mile in their shoes. You really cannot judge a person until you've sat down with them in their country and you've had food. And that's the thing that I think is the most important lesson we can push to the next generation.'

Yet much education seems to limit rather than unleash curiosity. We need to change that.

How do we learn curiosity? And how do we teach it?

We can start by learning about our relationship with curiosity over the centuries, and the eclectic cast of history's most ingenious people, who have helped us move from cave paintings to driverless cars. To be more curious ourselves, we need to understand how and why these breakthroughs came about, through a basic grounding in this history of human discovery.

What are the key moments in our collective story of innovation?

Some of our most ingenious ancestors were those who first walked and talked. As our ape-like ancestors evolved in East Africa they began to stand on two legs and use flint tools. By about two million years ago, these early humans had also learned how to make fire to cook, keep warm and protect themselves against attacks from animals. A hundred thousand years ago, we began to develop language to communicate, and started to express our imagination in cave paintings and carvings. Our curiosity was no longer just about survival.

As humans evolved, we worked out that certain grasses could be grown for food, and started to settle down to do that. We used metal, invented tools such as the plough, and domesticated animals. We developed early forms of writing to keep records of trade.

Irrigation (from roughly five thousand years ago) allowed early kingdoms such as those in Egypt, Mesopotamia and North China to produce surpluses and settle populations near rivers such as the Nile, Tigris, Euphrates, Indus and Yellow. Three and a half thousand years ago, our Bronze Age ancestors invented the wheel and axle combination that enabled us to move produce and eventually ourselves around.

Steam power became our next great discovery. For the first time, people had a source of power that was not dependent on nature. The First Industrial Revolution transformed production, standards of living and transport. From steam onwards, it becomes easier to discern the role of individuals in the process of scientific discovery and human development. Light bulbs literally came on with the discovery and mass use of electricity.

The ability to invent at pace and in company was transformed by telecommunications, with Samuel Morse's telegraph machine of the 1830s leading rapidly to swift transatlantic communication. In 1876, Alexander Graham Bell, who had come to America to teach the deaf, found a way to transmit speech electronically with the invention of the telephone. The first-ever call on a handheld mobile phone was made by a Motorola employee, Martin Cooper, in 1973. His handset weighed two kilograms. There are now more mobile phone subscriptions than there are humans on the planet.

Meanwhile, our access to information was transformed by an eclectic cast of nineteenth- and twentieth-century Europeans, Japanese and Americans whose tinkerings led to John Logie Baird's television system. Twenty million people watched Queen Elizabeth II's coronation in 1953. Three hundred million watched her sister's wedding in 1960. One billion watched her daughter-in-law's wedding in 1981. Two billion watched her daughter-in-law's funeral in 1997. A reality television presenter became US president in 2016.

For the vast majority of human history, infections killed more people than the wounds themselves. Human development was held up by disease, even while the efforts to counter it absorbed medics and scientists. But when British bacteriologist Alexander Fleming discovered penicillin in 1929, and armies tested it at mass scale in the Second World War, it enabled another leap forward in our collective story: the era of antibiotics. Smallpox, which had killed 300–500 million people during the twentieth century, was eradicated through a massive postwar cooperative effort. Double Nobel winner Marie Curie was followed by James Watson and Francis Crick in understanding our DNA and launching modern molecular biology. The subsequent Human Genome Project, to map all the genes of our species, has been one of the great collaborative international innovative efforts: American-funded, but with research centres in seven countries.

Computers have accelerated our ability to solve problems together. Grace Hopper designed a five-ton, room-sized computer at Harvard in 1944, from which she had regularly to remove real bugs. Since then the number of modern transistors in integrated circuits has almost doubled every two years, meaning that computing power has since been able to grow greater than exponentially. Spurred on by passion, rivalry and enormous amounts of potential money, Steve Jobs and Steve Wozniak founded Apple Computer in 1975, determined to make a computer small enough for the home. IBM came to the party, and Bill Gates' Microsoft provided the operating system. The 1984 Apple Macintosh model evolved into laptops, tablets and smartphones. Everything changed again, creating the basis for the extraordinary pace of technological innovation – from Mars missions to driverless cars – of the years since.

These eclectic examples of human ingenuity share three features that can help us understand how we can build an environment that helps humans to be more curious in the future. First, the technology was hugely underestimated at the time. Many of the inventors and dreamers who contributed to these innovations were dismissed as cranks by their contemporaries: only in the second half of my scamper through their history do we start to learn their names. Many died frustrated, penniless and forgotten.

So, human evolution advances one crank at a time. But our survival depends on finding and creating more space for the individuals with the passion, expertise, time, resources, patience and eccentricity to labour at the process of scientific discovery. You need your Marie Curies, Alexander Flemings and Grace Hoppers. All these examples of extraordinary ingenuity required a human spark, a moment of genius. I like to think that they also required the rest of us to adapt and adjust, playing our small part in the process. Are our current education systems allowing space for the outliers to find their voice and be heard? How many geniuses are sidelined in class, or never make it there in the first place?

Practically, this means ensuring that classrooms and workplaces aren't stifling the non-conformists. It means ensuring that we don't create assessments or promotion requirements that always prize vanilla uniformity over quirkiness. We need to allow ourselves to think the unthinkable sometimes, and to nurture the individuals who do little else.

Second, much innovation is about development, disaster or disruption. Development fuels ingenuity as part of a much broader pattern of human self-improvement. Each new life-changing and economy-transforming technological change draws time, money and energy from those that preceded it. As populations become wealthier, they have more spare labour and more time for education. Governments are more likely to protect intellectual property or fund research. As we have seen, America's head start as a result of the Second Industrial Revolution set it up for greater success in the Third Industrial Revolution. The wealth of the car industry helped create the foundations for Silicon Valley: the road to San Francisco passes through Detroit.

Disaster fuels ingenuity when crisis or conflict forces humans to try new things: necessity is the mother of invention. The stumbling early human *Sahelanthropus*, our first ancestor to stand up, was probably trying to run from danger. Ingenuity often comes from the survival incentive, such as at times of war or a shortage of resources. Imagine the motivation of the Kampala graduate who, in a country where up to 27,000 children die every year from pneumonia, designed a jacket that can detect the disease.

The demands of the First World War alone generated several examples that continue to change our lives. Canned food became a lifeline. Tea bags were initially 'tea bombs', developed to supply soldiers with their essential brew. Daylight saving was originally an attempt by Germany and Austria to conserve fuel. The trench coat needs no explanation. Joseph Pilates invented his exercise method while a prisoner of war. War has even generated innovation in sweets. M&Ms were invented to give soldiers a quick energy boost, the pellets of chocolate coated in hard candy to prevent the precious chocolate from melting in their hands.

Perhaps we will see a fresh wave of innovation as a result of the adaptation to the lockdowns of 2020 and 2021? The challenges of the pandemic accelerated updates to how we interact with the world. They forced us to find new ways to communicate and collaborate, to draw strength and camaraderie from human contact even when we were unable to meet in person.

Disruption also fuels ingenuity, often when individuals or companies see a chance to displace an earlier technology that previously seemed unassailable. Think of the way that the credit card or instant noodle swiftly dominated a market. In 1976 Kodak accounted for 90 per cent of film sales and 85 per cent of camera sales in America. The first digital camera was invented in 1975 by Steven Sasson, an engineer at the company. It was then the size of a toaster, but thirty-six years later Kodak filed for bankruptcy, a victim of disruption – the ability to place digital cameras in our smartphones – that had initially come from within.

So development, disaster and disruption drive creativity. We need to find more ways to recreate the urgency that is unleashed by these forces. Practically, we need to ensure as a society that we are investing some of the benefits of the Fourth Industrial Revolution in the skills and knowledge we will need for the Fifth Industrial Revolution. This will mean scaling up the finance that is available for research and innovation.

Sometimes of course that moment for innovation can be the result of a combination of these factors. Penicillin became significant thanks to the coming together of research, war, government investment, and cooperation between British and American scientists, plus the commercial incentive that could rapidly reduce production costs. Alexander Fleming's name goes in the history book and on the Nobel Prize. But it was Oxford professors Howard Florey and Ernst Chain who turned his laboratory curiosity into a life-saving drug.

There is a third practical lesson from the history of curiosity: these innovations did not just disrupt existing industries, they disrupted society. The television, phone and internet changed not just the way that we receive information but how we understand and interact with the world. The countries that mastered the new technology – in those examples the USA – dominated the next century. And those that failed to do so fell behind. Our survival prospects for the coming century will be shaped to a large extent by whether we are part of a society that masters or is mastered by the next advances in technology. So we need to work harder to observe, understand and anticipate how technology is changing how we organise and live our lives. We'll come back to this survival skill in Chapter 8.

As well as these practical lessons for how we renew society, there is much we can do as individuals to develop the survival skill of curiosity. We may not discover the next life-saving vaccine. But we can improve our individual and collective survival chances. The application of our curiosity could be anything from spotting a potential gap in the markets to finding a new way to recycle plastic bottles.

As with the other survival skills, we can observe, learn, practise and teach curiosity. When we think of curious people, what is it that makes them different, in the way they think and in what they do? Leonardo da Vinci must have been one of the most curious people in history. That curiosity led him to great art, and to discoveries in anatomy, physics, biology and engineering. Yet he never received a formal education beyond basic reading, writing, flying and arithmetic. His genius came in questioning the world around him, combined with an ability to apply intense focus to the answers he found.

But this wasn't simply scattergun curiosity. Leonardo had a plan. In *Da Vinci's Ghost: Genius, Obsession, and How Leonardo Created the World in His Own Image*, Toby Lester has translated a page from the artist's 1490s notebook, with his to-do list for one day. It includes his aspirations to: measure and draw Milan; find a decent book on its churches; get the Master of Arithmetic to show him how to square a triangle; ask a Florentine merchant how they go on ice at Flanders; ask a maestro how mortars are positioned on bastions by day or night; examine another maestro's crossbow; find an expert to explain how you repair a lock, canal and mill; and ask another about the measurement of the sun.

It makes my to-do list seem pretty parochial. But it shows us that curiosity is a choice: we can decide each day to set out to ask questions. It shows that we do not have to limit ourselves to one field: the most exciting discoveries are often in the intersection between subjects. All it takes is the patience to seek out and learn from experts. You no longer have to wander around Milan to do this: imagine what Leonardo would have made of YouTube.

Here are nine ways to be more like Leonardo, and to practise curiosity as a survival skill.

1) OBSERVE A CURIOUS PERSON

Tim Berners-Lee, the British software engineer who discovered the world wide web, is the most influential human alive. *Time* magazine's list of the 100 Most Important People of the Twentieth Century makes clear that 'he created a mass medium for the 21st century. The World Wide Web is Berners-Lee's alone. He designed it. He loosed it on the world. And he, more than anyone else, has fought to keep it open, non-proprietary and free.' It took radio thirty-eight years, television thirteen and the web four to reach fifty million users. Since then the internet has changed how we work, live, trade, consume information and entertainment, network, mate and learn, with billions yet to come online. Yet when he featured in the brilliant London 2012 Olympics opening ceremony, the modest and rarely seen Sir Tim needed a caption to explain who he was.

When I asked Tim to describe the process of ingenuity that led to his extraordinary breakthrough, he was characteristically self-effacing. 'I guess I was just in the right place at the right time.' But when pressed to dig deeper, it is clear that, as with Melanie Perkins, the key was a mixture of patience, collaboration and eureka moments. His came thanks to the scientists who visited CERN, the European Centre for Nuclear Research where Tim was based, and needed to store and exchange their results. This practical requirement drove the research into how it could be achieved. 'Most of the technology involved in the web was already designed, already out there. I just had to take this mixture of existing ideas and find the right way to connect them.'

The result of what Tim calls 'this act of desperation' was info.cern.ch, the world's first-ever website and web server. Visitors could learn more about the project, how to search the web or create a webpage. On the surface it was a pretty mundane moment. I wonder how many of those who visited the page in those early days imagined that the instant connectivity it established would top a list of cultural moments that shaped the modern world.

For all of us trying to develop our own curiosity or encourage it in young people, where did it come from for Tim? Alongside voracious reading, he credits the example of his parents. His mother and father were mathematicians who worked on the first commercial computers. Tim learned about electronics through playing with model trains. At university, he made a computer out of an old television, which he bought from a repair shop.

The rest was being in the right place at the right time, having the right dream and putting in dedicated effort. 'I hope my achievements can be a lesson to all dreamers,' he says. 'Dreams can come true when you try hard enough.' Barack Obama wrote something similar in the book of advice I keep for my eldest son (see Chapter 4). 'Dream big dreams, yes – but then go and work for them.'

2) PRACTISE PROBLEM SOLVING

Everybody has the capacity to think differently. Just as with learning any new skill, exposing ourselves to problems and challenges encourages us to think creatively about how best to solve them. As English author Neil Gaiman says, 'The imagination is like a muscle. If it is not exercised, it atrophies.' Curiosity is not unique to people we consider to be geniuses. We all have it.

Practice doesn't mean memorising stuff. Surveying the increased access to information in his own era, Albert Einstein counselled that we should never waste brain space memorising information we could look up in books. Now we need not remember information that we can access on our smartphone. But we need to know how to find it, understand it and use it.

In other areas of life – learning an instrument, starting a new job, improving our fitness – we take it for granted that we'll need to put in the time and effort. Why not do the same with the way we develop our brains to wrestle with problems? If human history is the story of how we have tinkered with the world around us, maybe it is time to get the screwdriver out and fix that wobbly shelf.

3) JOIN THE DOTS

The key to so much of our creativity is crossover, the ability to make links between ideas and disciplines. The success of universities like Oxford is that the ideas fuse and develop in extraordinary ways not just because a chemist is in the same lab as the greatest chemists on the planet, but also because they sit at lunch with the greatest historians and anthropologists on the planet. The seed of a new idea comes so often from seeing it from a completely new angle.

We've seen how Tim Berners-Lee's moment of genius came in fusing what was already out there. Biographer Walter Isaacson has written that Steve Jobs 'connected the humanities to the sciences, creativity to technology, arts to engineering.

There were greater technologists (Wozniak, Gates), and certainly better designers and artists. But no one else in our era could better firewire together poetry and processors in a way that jolted innovation.'

As Jobs put it in a 2005 Stanford University commencement address, 'You can't connect the dots looking forward; you can only connect them looking backward. So you have to trust that the dots will somehow connect in your future. You have to trust in something – your gut, destiny, life, karma, whatever.'

If cross-pollination is the root of many breakthrough ideas, we must get better at it. Take an article about a different field that strikes you as interesting and break it down into key ideas in your own words. Then try to apply each of those ideas to a new, different field. For instance, an article about how the legal sector is being disrupted by technology will throw up a series of interesting examples of both the disruption and the creative response. What would these mean in your field? Could you write the new version of the piece?

Another great way of joining the dots is to take the equivalent of a random walk. Look up a fact about a character, place or idea. See where it takes you. Keep seeking connections to the problem you are trying to solve. For example, when trying to find the answer to a problem about innovation as a result of the lockdowns of 2020 and 2021, explore innovation in previous pandemics or crises. Where are the dots? What can we learn?

4) SEEK OUT FASCINATION

Five times stronger than steel, Kevlar is used in hundreds of products, from bike tyres and tennis rackets to frying pans and bulletproof vests. It was invented by Polish-American chemist Stephanie Kwolek in another moment of luck, tinkering and seeking out fascination. In anticipation of petrol shortages, her team at the chemical company DuPont were looking to replace the steel in tyres with something lighter weight, to reduce fuel demands. She concluded that 'all sorts of things can happen when you're open to new ideas and playing around with things'.

Read. Read about ideas. Read about people. How did they become great at what they do? Learn from that. Try things you don't know how to do. Learn from that. Bill Gates famously reads fifty books a year. Armed with that knowledge, we can seek those connections. We don't have to be involved in world-changing ideas to do this.

Another great way to develop this skill is to put aside that Amazon order and visit a physical bookstore or library and browse. The lockdowns made this harder, but there is no substitute for doing this. Follow the advice of the economist John Maynard Keynes: 'A bookshop is not like a railway booking-office which one approaches knowing what one wants. One should enter it vaguely, almost in a dream, and allow what is there freely to attract and influence the eye. To walk the rounds of the bookshops, dipping in as curiosity dictates, should be an afternoon's entertainment.'

Writer Josh Kaufman's mission is to help people 'upgrade the software in their brains' by becoming more effective at reading and distilling information. These are the skills that helped him diagnose his own mystery illness. On any given day he'll spend up to 75 per cent of his time reading. 'The more reading I do, the more I've discovered that people aren't typically paying attention to a huge corpus of pre-existing knowledge, often decades-old,' he says. 'Productivity isn't about checking everything off your to-do list – that's an unwinnable race. It's about deciding what's important to you, your values and priorities, then directing your time and energy at those things as much as possible.'

5) PLAY MORE

Albert Einstein had a turbulent relationship with his son, yet (as with Maya Angelou) this letter to the boy captures this important and replicable attitude to learning.

Yesterday I received your dear letter and was very happy with it. I am very pleased that you find joy with the piano. This and carpentry are in my opinion for your age the best pursuits, better even than school. Because those are things which fit a young person such as you very well. Mainly play the things on the piano which please you, even if the teacher does not assign those. That is the way to learn the most, that when you are doing something with such enjoyment that you don't notice that the time passes. I am sometimes so wrapped up in my work that I forget about the noon meal.

Creativity is not something that we or our kids should do in our spare time, like an after-school arts club. It should be a key part of everything we do. Giving ourselves the time to play and explore is vital to our curiosity. When I visited Tech Will Save Us co-founders Bethany Koby and Daniel Hirschmann in their lab in London, they showed me how kids themselves develop their learning kits. 'What's more extraordinary than watching a child solve challenges?'

The company LEGO has long recognised this. Its vice president Ryan Gawn told me that the 2020 lockdown made unstructured play even more essential. 'Families who do well play together, and are often happier, have better connections to their children and are reminded about what is important in life.'

So if nothing else, get some toys out. Protecting play at home and in school increases family happiness, wellbeing, development of life skills and better learning among children. Play develops communication skills, abstract thought, self-regulation and more adaptive, flexible, creative thinking. LEGO has suggested five types of essential play for children, all of which our NYU research shows can apply to adults too:

- Physical activities, such as jumping, climbing, dancing, skipping, bike riding and ball play; fine-motor practice such as sewing, colouring, cutting, manipulating action toys and construction toys; and 'rough-and-tumble' play.
- Pretend play, such as make-believe and role-play, which develops reasoning skills, social development and creativity. If a child pretends to be a fictional character (like a superhero), they become more immersed and stay focused on a task.
- Play with objects. This begins early, with behaviours such as basic manipulation, rotating objects while looking, hitting and dropping. It progresses to arranging and constructing objects as toddlers and then develops into sorting and classifying until the more complex building, making and constructing of larger objects.
- Symbolic play. This begins when children first start communicating and progresses to include spoken language, mark making, numbers and music, by using the first signs of symbols and representations.
- Games with rules, including sports and board games.

We can sometimes suppress those urges to dress up, sing kara-oke, play with our kids' LEGO sets, do a jigsaw puzzle, or take on an epic game of Monopoly or Risk. Yet these are precisely the carefree activities that can help us to develop the survival skills we need.

6) CREATE SPACE FOR LUNACY

The apple falling on Newton's head. Archimedes and his bath-time eureka moment. Melanie Perkins on her mum's couch. Many of history's most ingenious revelations came from a single thought, no matter how strange or unlikely it might have seemed. But they all required the space to think, a conscious decluttering of the mind.

How good does it sound to create space for serendipity and peace? Yet how rarely do we actually do that? During the pandemic lockdown, most of us imagined that the silver lining would be greater freedom to rest and think. I fear I did even less of it. Social media and the press of news, work and life have filled huge amounts of the time in which we used to do nothing but look out the window. In my experience, doing nothing at all is becoming harder and harder to achieve.

'Here's to the crazy ones,' reads the Rob Siltanen quote in Apple's 'Think Different' ad campaign. 'The round pegs in the square holes. The ones who see things differently. You can quote them, disagree with them, glorify or vilify them. About the only thing you can't do is ignore them. Because they change things. They push the human race forward. Because the people who are crazy enough to think they can change the world, are the ones who do.' Given that humanity advances one crank at a time, maybe we can also allow ourselves to think crazy thoughts.

The mad ideas are sometimes the ones that cut through. And the people who succeed in realising these revolutionary advancements do so because they let themselves believe that a crazy idea *could* be a reality, even if they can't comprehend how: Elon Musk thought both Tesla and SpaceX would fail at the beginning.

In *Loonshots: How to Nurture the Crazy Ideas that Win Wars, Cure Diseases and Transform Industries*, Safi Bahcall has identified two features of scientific breakthroughs: serendipity and genius. Successful research labs like Bell Laboratories (currently sitting on nine Nobel Prizes and four Turing Awards) aim to combine the work of the artist and the soldier. The artist finds the crazy ingenuity and the soldier keeps the money coming in to fund the research. Much ingenuity requires the restless individual genius who can elevate an idea through brilliance and the right combination of patience and impatience. 'I will always choose a lazy person to do a difficult job,' Bill Gates is reported to have said, 'because a lazy person will find an easy way to do it.'

We need to allow time for the mad ideas and patience for the means to deliver them. Sometimes that might just mean allowing more space to the misfit thinker or encouragement to the lonely genius inside us all.

7) FIND COLLABORATORS AND COMPETITORS

This chapter has shown that for innovation, you need the cross-fertilisation of ideas, research and ingenuity. An invention or breakthrough often triggers others. It is unfair to single out individual inventors of steam power, electricity or computing technology, given the way in which so many people contributed to those discoveries. The wheel would not have been invented had the copper chisel that could make the axle sufficiently smooth not come first. James Watt's late eighteenth-century steam engine would not have been possible without Thomas Newcomen's 1712 prototype, which in turn would not have been possible without the experiments in steam pumps by Jerónimo de Ayanz y Beaumont, in steam turbines by Taqi al-Din or in steam cannon by our role model of curiosity, Leonardo da Vinci. Scientists could not have developed the first laser in 1960 without Einstein's work on light amplification fifty years earlier. Michael Faraday's electric motor would not have been possible without the research that had been carried out by a succession of figures, from polymath diplomat Benjamin Franklin in the eighteenth century to Alessandro Volta, inventor of the electric battery.

The same is true of whole fields, such as nanoscience and nanotechnology: the study of extremely small things. 'Nano' means one-billionth: there are 25,400,000 nanometres in one inch. The technology is now used in food science, water purification and space technology. It could only have happened as a result of a rolling team effort – indeed that rolling team effort, that collective tinkering, is the story of human ingenuity.

In almost all cases of modern discovery, humans have been working away separately, but with an increasing awareness of what others in the field have done and are doing. In the past, hubs such as market squares, trade routes, Parisian salons or Silicon Valley often provided the crucial mix of people and ideas for innovation to thrive. But this ability to co-create has been accelerated hugely by twenty-first-century connectivity: the hubs are now online.

We have seen how cooperation was as essential as competition to almost all the key breakthroughs in human ingenuity.

Steve Jobs designed the buildings he worked in to make sure that collaborative encounters happened: 'If a building doesn't encourage that, you'll lose a lot of innovation and the magic that's sparked by serendipity. So we make people get out of their offices and mingle in the central atrium with people they might not otherwise see.' Curious people surround themselves with those who challenge and inspire them.

It helps to identify the people we feel most energised and excited to work with, and work more with them. Or at least to identify people who leave us feeling drained and unexcited. And work less with them.

And usually a bit of competition helps. On the first regular broadcast of television, in November 1936, two systems competed: Marconi-EMI's 405-line system and Baird's 240-line intermediate film system. Marconi eventually won. Or take electricity. In the late nineteenth century, fuelled by competitive zeal, men such as Thomas Edison, Nikola Tesla and George Westinghouse made it commercially available and part of our lives. One of its key components, the MOSFET (metal-oxide-semiconductor field-effect transistor), has quietly become the most widely manufactured device in history, yet most of us are unaware of it.

8) ASK RELENTLESS QUESTIONS

At a stirring NYU Abu Dhabi graduation ceremony, my former student Dubai Abulhoul, who was voted Young Arab of the Year in 2016, said in her welcome speech that the most important thing the students had learned was not the right answers but the right questions: this allowed them to colour outside the lines. That is a mindset that should inspire and encourage us all.

other any question. If you were a colour, what would you
? What would you do if you were president?
Here are my eight questions to ask young people:

1. What did you learn today?
2. What excited you this week?
3. What's your story?
4. What are you reading?
5. What's the best thing that's happened to you this
   year?
6. Who would you most like to meet and why?
7. What problem do you wish you could solve?
8. If you could ask me anything, what would it be?

You'll be amused, inspired and challenged by the responses.
You will learn a huge amount. And most importantly, you will
have helped our collective survival chances by moulding a
curious brain and by increasing our herd immunity to incuriosity. 'Every child is an artist,' recognised Picasso. 'The
problem is how to remain an artist once we grow up.'

If you don't have a young person to talk to, talk to a
stranger. Find people who are interesting. Maybe start out by
aiming for a conversation with one person a week. Or just
putting the phone down for long enough so that someone
talks to you. Somewhere in the last two decades, especially in
cities, we lost that ability for spontaneous interactions. Yet the
way these can build community and help ideas move is a vital
part of our collective survival skillset.

9) SAY 'I DON'T KNOW'

Three words that seem to be heard less and less in modern discourse. Think how unusual it is to hear a politician admit to not having all the answers. Think how rarely we hear someone saying that they are still making up their mind about an issue. Or that they need more time before joining a campaign or sharing a post on social media. Evolutionary psychology suggests that men are more likely than women to deceive to bolster their status and influence. Yet the ability to say 'I don't know' is actually a sign of curiosity. 'Curious people aren't afraid to admit when they don't have an answer,' says LeeAnn Renninger, co-author of *Surprise: Embrace the Unpredictable and Engineer the Unexpected*. Practise saying less. Listen hard, and listen with empathy.

We can also take this questioning curiosity into our wider lives. As our favourite Stoic emperor Epictetus tells us, 'We have two ears and one mouth so that we can listen twice as much as we speak.' It may be the serendipitous find that sparks the idea. Google founder Larry Page described the 'perfect search engine' as one that would 'understand exactly what I mean and give me back exactly what I want. But what if I don't know what I want?'

If we nurture this kind of curiosity we increase our collective chances of making the breakthroughs we can't yet imagine but will find we cannot live without. As I write, humans are making the blind see, through inventing artificial corneas and irises. They are making the lame walk, through inventing artificial limbs that don't just replace but enhance. They are extending our lives, through pushing back the boundaries of medical research. They are bringing internet access to every corner of the globe, connecting humanity in a way that was unimaginable even a decade ago.

These are all miracles in themselves. And yet they no longer strike us as miraculous. We have made the impossible possible and the extraordinary ordinary. That should give us hope. Back to Melanie Perkins in Perth and her Canva business conquering the world. Like so many of the innovators in this chapter, she learned as much as she could about her field. And then she lay on her sofa and created space and time for the crazy idea, to join those dots. We can do all that too.

So we have taken back control of our time, and the parts of our life that we can control. We have found new ways to keep a sense of wonder and fascination, to ask questions and solve problems. What do we do with this time and with this curiosity?

3

How to Find Purpose

I want you to play another octave.
John Sexton, President Emeritus,
New York University

The poet and essayist Molly McCully Brown has lived her entire life with cerebral palsy, which restricts her movement and causes her near constant pain. Despite, or maybe because of this, she has travelled widely, writing powerfully about the intersection between what it means to be human and 'the places I've taken my body'. When I asked what purpose meant to her, she said, 'It means not being defined by how people want you to be. I'm not pretending that I don't despair some-times about how hard it can be to live the life I want, to go to the places I want. But it is then that you reach deeper. And find your real self.'

This search for our real selves has been part of the human story. It has often determined whether or not people made the impact they aspired to. We have seen how the search for purpose can help us to take back more control and become more curious. We can picture people in our lives who have purpose. What is their secret? How can we learn it and teach it?

Purpose does not mean a sense of complete conviction, an unassailable confidence. Many of those with the strongest sense of purpose are also those with a healthy dose of self-doubt too. But they have gone through a process of asking themselves who they are and reorienting their lives towards that North Star.

Perhaps the starting point is in understanding the way that others define us. When I challenged Filippo Grandi, who runs the UN Refugee Agency, to define purpose he asked me to imagine the experience of being a refugee. The daily fear, indignity, rejection and hopelessness. 'That's where their sense of purpose comes from. They know what happens when they let it go. But often their courage comes from facing down the label of refugee – not letting it define you.'

He described Lubab al-Quraishi, an Iraqi refugee who is now on the Covid-19 frontlines as a medical worker in New Jersey. She had been a pathologist in Iraq until being targeted because her brother worked with the US military. In the US she sold burgers to keep her family alive, until in 2020 the pandemic led to a demand for pathologists, and the chance to pursue her craft again.

Not all refugee stories end with such hope, but her story is not unique. For Filippo, there is much that we can learn from refugees about our common humanity, but the courage to face down that label is key: 'We have a tendency to want to classify people by nationality, race, gender, social or economic status. On those small boats in the ocean aren't one ethnic or social group, but individuals with different dreams, backgrounds, aspirations. Don't let people tell you who you are.'

Is our sense of purpose defined by other people? Do we allow society, family, career or economic status to tell people who we are?

In the study of geopolitics, soft power is a growing field. It captures a sense of whether a country is magnetic, and the values it projects. Yet most nation-branding exercises fail to take proper account of what the rest of the world already thinks. A country such as the UK or France that is aiming to be seen as vibrant, diverse and modern often fails to square this with a history of empire and protocol. Or a country that wants to be seen as a pioneer on the climate crisis may fail to reconcile this with the fact that its wealth came from fossil fuels. When coaching ambassadors, as I now do, I get them to write down the three words they want people to associate with their country, and we then look at the three words that studies show people do actually associate with it. The differences are often striking.

To assess whether we are succeeding or failing in the attempt to define ourselves as individuals, we can try the same exercise. Write down three aspects of your life or personality that are vital to how you see yourself in the world. I might write: 'I'm a good mentor.' 'I thrive on the adventure of travel.' 'I inspire others through my leadership and example.' Then write down the three things you believe people actually think. Or, if you're feeling really brave, ask them. You might find that what they truly think is that you talk too much about yourself, you take a week in a resort once a year, and you throw your weight around the office.

Recognising the extent to which these two lists vary is a good starting point for redefining who and how you are in the world. At a time when it is easy to feel buffeted, we can develop this advanced survival skill by identifying what really matters to us. As my friend John Casson said, we become what we pay attention to. Write it down. Share it with someone you trust. Hold yourself accountable.

I think that the next stage to finding purpose is to put together your personal manifesto. What are you aiming for? What promises do you make to yourself and those you care about on the life you aspire to lead?

I tried to write out my five personal objectives for the future. It is much, much harder than it sounds. I wrestled with many versions. But here – in case it helps – is where I ended up.

- An adventure-filled, passionate, supportive and empowering marriage.
- Two curious, kind, brave boys. And to be a good ancestor.
- To own a special piece of land.
- To leave a decent eulogy as a global citizen of resilience, humanity and energy.
- To learn and teach something every day.

It is an imperfect list. It also regularly serves to remind me where I'm failing. But that is the point. When I face major life decisions, I have something to hold fast to, a North Star to orient myself towards. We might think it is obvious to ourselves and everyone around us what is in our manifesto. Maybe you are one of the lucky and blessed people for whom that is true. But for me it was only when I tried to write it down that I realised how difficult it was for me to see it, let alone live it.

Another way to help guide the process is to try to visualise your life in 2050. Write a sentence on each of:

- Family
- Health

- Work
- Purpose
- Wealth
- Learning

For each of these headings, try to complete the following statements.

- My hope is that ...
- One practical step towards this that I can take now is ...
- One essential ally I have is ...
- I will know I am making progress towards my hope when ...
- The main challenge/obstacle within my control is ...
- The main challenge/obstacle beyond my control – that I need to understand better – is ...

The answers may give you renewed confidence that you are on the right track. They might lead you to make fundamental changes. Or most likely, they may nudge you to make small, subtle changes. But once considered, those answers will be there in the background, giving you some scaffolding when it all feels a bit fragile, quietly helping you to navigate the turbulent waters of the 2020s.

You are most likely to do this in your professional life. For my foundation, which aims to help people do good things in public life, we spent days thinking about our purpose. We realised that we wanted to back innovative diplomacy in a time of closed politics; the spread of opportunity in a time of inequality; and creative education in a time of automation. We wanted to be pioneers, architects and citizens working for opportunity, creativity and coexistence.

This book is one result of that discovery.

Once your sense of purpose is clear, it is easier to avoid letting the process suck the oxygen out of the substance. It is easier to be about a project than a position. It is more probable that others will support and follow you: people don't just want to know what you do but why you do it.

And it is likelier that you will have the resilience for when it is not going to plan and the humility for when it is. It is easier sometimes to simply be, not just do. 'Most men lead lives of quiet desperation,' wrote the reclusive American writer Henry David Thoreau, 'and go to their graves with their song unsung.'

So what's your song? And is it unsung?

There is an advanced version of this exercise to define purpose, but it is demanding and emotionally draining. It is to try to write your own eulogy.

The *New York Times* columnist David Brooks has written powerfully about the difference between resumé (or CV) virtues and eulogy virtues. His original article on the subject had as much impact on me as anything I have ever read. 'The resumé virtues are the skills you bring to the marketplace,' he writes. 'The eulogy virtues are the ones that are talked about at your funeral – whether you were kind, brave, honest or faithful. Were you capable of deep love? We all know that the eulogy virtues are more important than the resumé ones. But our culture and our educational systems spend more time teaching the skills and strategies you need for career success than the qualities you need to radiate that sort of inner light. Many of us are clearer on how to build an external career than on how to build inner character.'

Another way of thinking about this is to reflect on the first line of your obituary, and the first line of your eulogy. The former is what people who don't know you think you did.

The latter is what people who know you think you are. Both are part of finding purpose. And the closer they are, the more you will be living a life that is authentic to that purpose.

Just as thinking about your purpose can help your professional and life goals, thinking about how you really want to be remembered can have a profound impact on how you live your life. It can help you to open your mind and open your heart.

The process of defining and redefining your purpose depends on something that is also key to our curiosity. It can be seen in the reply of Google DeepMind founder Mustafa Suleyman to a request for the most important skill young people can learn in order to interact with technology. 'My biggest piece of advice is,' said Suleyman, 'what does it mean to ask good questions. Because effectively that's what you will be doing with a machine as you grow older. So being the expert Googler almost, are you better at searching than your peers, can you find the right answer among hundreds of other answers? That's almost a better skill than knowing the answer.'

So it comes down to asking the right questions.

As part of the NYU research that we conducted on the future of learning, we developed and tested a set of questions that can be used to open up debate and challenge within organisations. Here are twenty questions I've gone on to use with governments, NGOs and corporations to help them challenge themselves about their purpose.

Mission

1. What do you really do for people that no one else does? What guides everything you do?
2. Do your people know why they are there?

3. And what do they talk about in the lift?
4. What do the pictures on your walls say about your organisation?
5. Are you fighting the last war?
6. What are the outcomes that define success or failure?

People

7. Are you playing Minecraft or Tetris: looking for collaborative ideas or top-down diktats?
8. What were the biggest risks you took in the last decade?
9. How genuinely representative are you of those you serve? How accountable are you and your people?
10. Are you honest with staff about the deal on offer?
11. What are your distinct characteristics? What skills do you really reward?
12. What does your organisation look and feel like on your best day?

Means

13. What would a 'behind the scenes' documentary reveal?
14. Are you gaining or losing power?
15. If you were occupied by a bunch of smart, values-driven, tech-savvy millennials, what would they improve?
16. Do you have a decent system to manage your network and knowledge?
17. Are you a castle or a connector: do you hoard power behind a wall or use it to convene and open up?

18. What have you learned from your opponents?
19. Are you fast enough?
20. Are you slow enough?

In the introduction I described the ten megatrends that we need to anticipate in the next three decades. If we apply the lessons of Chapter 1 on anticipating the future to those, I think that there are three underlying shifts in our collective consciousness that can both help us define our purpose and have the greatest potential to derail our aspirations if we fail to understand them. These are the rise of distrust, an increasing awareness of inequality and the fear of technology. Get on the right side of those trends and we have a greater chance of survival, as individuals and communities. Fail to do so and we will continue to feel knocked off course, flailing and overwhelmed. Each can both help us define our purpose and influence whether we can live by it.

Trust is plummeting. In 2020, the US public relations consultancy Edelman – which measures trust each year – found that none of government, business, NGOs or the media is trusted by more than half of the population. In previous surveys, it had found that economic growth increased trust. That was no longer the case: more than half the 2020 respondents said capitalism was doing more harm than good. The gaps in trust are increasing rapidly between wealthier, more educated news consumers and the mass of the population. Government, more than any other institution, was seen as least fair: 57 per cent said government serves the interest of only the few, while 83 per cent of employees feared losing their job, attributing it to the gig economy, a looming recession, lack of skills, cheaper foreign competitors, immigrants who will work for less, automation, or jobs being moved to other countries. Fewer than a third of people believed business would pay decent wages and provide retraining for those whose jobs are threatened by automation.

This loss of trust is corrosive. The Iraq war reduced trust in the foreign policy establishment, which was one reason why governments couldn't bring the public with them for military action to prevent Bashar al-Assad's atrocities in Syria. The MPs' expenses scandal, the financial crisis and EU misman-agement reduced trust in Westminster, the Square Mile and Brussels, contributing to fewer people entering politics, a growing suspicion of business, and Brexit. Donald Trump was a rejection of the establishment and mainstream media. But most worrying, public trust is plummeting not just in politi-cians, the media and the banks, but in teachers, doctors and the police. The smartphone has given us a sense of individual agency, and it has reduced our confidence in traditional authority.

What does this mean to a twenty-year-old in 2022? A healthy dose of scepticism could be replaced by a sense of fragility, a feeling of being unable to rely on anyone. This assumption of malign intent on the part of everybody in government, business or the media will make it harder to govern, to build businesses and to defend a robust and inde-pendent media.

The second big trend is the growth in the perception of inequality. There is a famous experiment by the primatologist Frans de Waal. In it, two capuchin monkeys are given cucum-bers as a reward for the same task. Then one of them is given a grape as its reward instead. The disadvantaged monkey sees this and tries to rip the cage apart. This is the oxygen on which extremists in all our societies rely. And if the economic trends described earlier in this book continue, we are right to worry. How do we create more jobs? How do we develop the right skills? How do we match the right people to the right roles?

How do we create more winners from all this change while better protecting those left behind? The case for reducing the gap between rich and poor is a humanitarian one, but it is also a matter of security: people need faith in their future and their society in order to buy in.

The third big trend we are living through is more existential. At our current pace, we will face as much technological change in the next century as we did in the previous forty-three. As I've described, it will rip through everything we think we know about society, the economy, the workplace and politics. Look at the impact of electricity, supercharging changes in how we live and work, and imagine that on steroids.

This will also create new tensions, new conflicts. If digital information is the twenty-first century's most precious resource, the battle for it will be as fierce as the battles for fire, axes, iron or steel. As those between libertarians and control freaks: people who argue that we should have complete freedom as humans and those who see the need for greater constraints. Between sharers and exploiters: those who see the advantage of technology to build together and those who see opportunities to destroy. Between those who want transparency – including many individuals, companies and governments – and those who want privacy (or, as its critics call it, secrecy) and the protection and hoarding of information. Between old and new sources of power. No wonder we feel uneasy at the prospect.

We all need to ask ourselves where we stand on those three trends, as individuals, organisations, communities and nations.

Are we gaining or losing trust? Does the way we spend our time add to or subtract from our bank of trust?

Are we increasing or decreasing inequality? Can we be confident that the choices we make are contributing to a fairer world?

Are we using technology or being used by it? Have we made the right adaptations to a world in which technological change at this pace is the norm?

If we can answer these questions honestly, and make the changes the answers point us towards, our survivability will be immediately increased.

Applying these questions to ourselves as individuals throws up tough choices. Am I working in a field that is gaining or losing trust? Do I build trust in the way I handle my relationships? Does my work or wider life contribute to inequality and injustice? Will future generations question my choices? Is technology helping me to live a better life, judged against what really matters to me?

How do we build the vital, powerful currency of trust at a time when trust is becoming harder to get and easier to lose? A huge amount of our personal impact will come from the small moments: those interactions we have with those we meet, live or work alongside. We build trust through our credibility, reliability and intimacy: what we know, how we communicate it and how we build rapport. But unlike any previous generation, much of the way we build trust in the future will also be digital. And, for better or worse, it will stay with us, as we move between professions and cultures. It is becoming harder to reinvent ourselves, or to wish away the social media blunder that we made a decade ago.

The cultivation of trust – online and offline – requires much greater attention than we currently give it. As a wise diplomat advised me on being an ambassador, don't worry so much about what they say to you when you are in the room, and worry more about what they say about you when you leave it. Psychologists who are experts on personal impact call this the primacy effect and the recency effect: formed by the initial impression of a person and the lasting memory of them. The initial impression is often realised within fifteen seconds, with more than half of the impact conveyed by our body language.

Of course, few people really want a world in which our online trust quota is assessed by whether we have enough Instagram followers or a blue tick on Twitter. And if we seek joy in those signs of recognition we will quickly become frustrated and restless. I am wary of an existence in which every human interaction is rated like an Uber driver and his or her passenger, with the results aggregated to give a life score. A combination of *The Truman Show* and *Big Brother*.

But I do think that there are practical ways to develop this idea of managing your personal brand, your bank of trust, over a lifetime.

First, do a few things really well that add value. That will help you to make yourself known – across fields and cultures. They will become the brush strokes – the lines that help to define you in people's eyes. You often see people doing this in their social media biographies. In just 280 characters, how do you describe yourself? And how do other people describe you? These can help shape your interactions online and offline. They can also help you shed what the politician and diarist Chris Mullin used to say was 'PA: pointless activity'.

Second, pay more attention to your life reputation. Without obsessing about every Facebook comment or negative media article, we can tend to the information that is out there on us whether we like it or not. And without obsessing about every grumpy exchange or bad commute, we can tend to the interactions we are having in real life every day.

Third, if nothing else, there is the most obvious but often neglected point on the accrual of trust. If we want to develop trust, avoid doing things that are not trustworthy.

Finally, own your mistakes. They make us who we are. I've screwed up countless times. I've learned to forgive myself. It makes sense to talk about your mistakes on your terms rather than let others unveil them on theirs. It is sobering to hear from so many people that they would choose not to enter public service because of the levels of scrutiny. The best leaders I've observed close up have a way of accepting and sharing their weaknesses that can be disarming and inspiring.

Part of finding purpose is to free ourselves from the fear of failure. How?

As we saw in the last chapter, the great pioneers of history made multiple mistakes along the way, yet nearly all accepted that this was simply part of the process. And crucially they all had the courage to learn from those mistakes. This attitude to failure and risk is a particular feature of entrepreneurship. Look at Microsoft creating the Xbox, the FedEx founder raising cash on the tables in Vegas, Elon Musk putting $165 million of his PayPal capital into SpaceX and Jeff Bezos giving up a highly paid Wall Street job to start Amazon from his garage. Again and again CEOs of very successful corporations have attributed their success to their failures. Failures created Henry Ford's Model T car. Walt Disney came up with Mickey Mouse after his first studio went bankrupt in 1923. Apple's then chief design officer Jonathan Ive said that Steve Jobs would often say to him, 'Hey, Jonny. Here's a dopey idea.' More often than not it would be a non-starter. 'Yet, every now and then, the idea would leave us both completely silent.' Facebook founder Mark Zuckerberg agrees: 'The biggest risk is not taking any risk ... in a world that's changing really quickly, the only strategy that is guaranteed to fail is not taking risks.'

Or, as Arianna Huffington, who defied her family members' cautions against failure and co-founded the *Huffington Post*, puts it, 'Failure is not the opposite of success – it is the stepping stone to success.'

J. K. Rowling wrote the first Harry Potter book while on welfare and while suffering from depression: the manuscript was rejected multiple times. She gave my son three pieces of advice: don't smoke, don't look down on people and don't fear failure. As T. S. Eliot wrote, 'Only those who will risk going too far can possibly find out how far one can go.'

Social entrepreneur Tony Bury told me he failed all his mock GCSEs and was then hospitalised. 'Everything I've achieved was from what that experience taught me,' he says. 'Human capital is capacity and capability: an individual's capacity is built by taking on challenges, which sometimes lead to failure. It is the failure, not the success, that builds human capacity. As you embark on your new journey you have to find a new network of friends; enter the storm. If you succeed you will have really learned and if you fail you learn a new perspective on life. As an entrepreneur you dip into that storm all the time, gaining capabilities and capacities to take on the world. It gives you resilience, gives you perseverance.'

I asked Tony, now a governor of a school that is actively trying to teach failure, how we can build this into our own lives without failing our exams and ending up in hospital.

'As you can imagine, there are real challenges around parents accepting their children will fail,' he says. 'There's a great piece of philosophy around this written by Joseph Campbell called a hero's journey, about how you as a human being allow yourself to have a call to change, and you step over the abyss into this new world of potential failure. Parents should benchmark schools on pastoral care before grades. How are they allowing that child to find their passion, or maybe a talent that was hidden somewhere that they never knew they had.'

The Lebanese adventurer Maxime Chaya has climbed the seven major peaks, cycled across the Empty Quarter and walked to both Poles, a unique set of achievements. Yet when I pressed him to explain which he learned most from, he said none of them: 'It was actually an earlier ascent of one of the mountains in the Himalayas. It was getting dark and the weather was turning. We had done all the preparations and this was the final window before the season ended and the route became impassable. I went up with four Spaniards, only three came down. The weather turned on us, and we had to decide whether the mission was more important than the lives of those on it. For the sake of those with me, I had to have the courage to turn back. We have to get past this fear of failing.'

When we read the biographies of figures such as Winston Churchill, the story rarely lingers on the periods after they failed. And yet those moments defined them. 'Failure Clubs' have become common not just in California startup circles but also in NGOs and the self-help sector. In 2008, Engineers Without Borders Canada launched a 'Failure Report' that has since been copied across the international development sector.

Yet many education systems penalise failure too heavily, stifling curiosity and creativity. A better option is to help children develop ways of problem solving through trial and error. In an experiment run by product designer Peter Skillman, small teams were given a few pieces of spaghetti, a length of tape and a marshmallow, and told they had eighteen minutes to build the tallest self-standing structure they could, with the marshmallow on top. Pre-schoolers were the most successful, and business school students the least. Skillman concluded that the kids were not afraid of failure: they just built a tower, watched it fall, and repeated this as many times as they could until they had a stable structure. By discouraging students from failing, we are discouraging one of the most effective means of learning. The author Angela Duckworth has pointed the way on grit and education: people with grit are more likely

Valerie Amos was the first black woman in the UK cabinet, and went on to run the UN's humanitarian operations before leading the School of Oriental and African Studies and now an Oxford college. When I asked her where her curiosity came from, she didn't hesitate. 'The dinner table. We arrived from Guyana when I was five. My parents always encouraged us to have debates and arguments at meal times. You had to come with new ideas and be ready to have them challenged. It taught me to think critically, to know how to defend my ideas, and which ideas to defend.' It was that ability to identify an argument and hold her corner that helped her succeed at the cabinet table, in the boardroom and in higher education. 'It is important that students bring a certain ragamuffin, barefoot irreverence to their studies,' said scientist Jacob Bronowski. 'They are not here to worship what is known, but to question.'

'The idea that a company's senior leaders have all the answers and can solve problems by themselves has gone completely by the wayside,' says consultant Ellen Kumata. 'You have to be rigorous: test your assumptions, don't take things at face value, don't go in with preconceived ideas that you're trying to prove.'

How do we learn to do this more ourselves? Kathy Taberner, co-founder of the Institute of Curiosity, says that curious people ask open questions. 'They stay away from questions that can be answered with a yes or no. This creates openness for the person who is being asked, and for the person who is asking.' I find few things more depressing than watching a family in a restaurant who are all on their phones: they are missing out on the most precious, finite moments to ask questions, to see where curiosity can take them. At home we ban only one word: bored. One of my favourite games with my youngest son is Spotlight, where we take it in turns to ask to overcome stress and use failure as a means to achieve their goals.

In the Philippines, one NGO is trying to plug that gap. TULA Learning focuses on character via what it calls 'missions', real-world challenges that learners attempt, based around themes like financial management, arts and culture. As students move through the missions at their own pace – the length of the programme entirely depends on the individual – there is a specific focus on risk, courage, drive, perseverance and potential.

When I asked neuroscientist Tara Swart how we should change education, she also highlighted the danger of teaching kids to always succeed. 'One of the fundamental things about education that I think needs to change is that the current model teaches us that failure is something very bad that we should be ashamed of,' she says. 'And it's often after school that people start learning by experimentation and learning from failure and getting much more comfortable with that. We shouldn't have to wait until we've finished school and you've started to explore more in life. We could be teaching those things in parallel with education: the feeling of experiencing something going wrong, the feeling that if you learn you actually become better and that actually you can always become better and that it should be about lifelong learning, formal and informal.'

Of course for every J. K. Rowling or Walt Disney, there are many who never get the break, even when they risk everything. This courage to take risks does not mean recklessness. American entrepreneur and inventor Mark Cuban has described his relationship with risk as adversarial. 'I hate risk. I'm terrified of it.' So he works ruthlessly to understand and mitigate risks with thorough preparation. 'I was relentless about learning.'

You can help to manage the fear of failure by compiling your own risk register. Identify the top-ten risks to you as an individual or family. Work out for each how much control you have over managing it. For each, take an estimate of its probability, identify the factors that might trigger it, and the wider trends that make it more or less likely. Then rank the risks, based on probability and threat. For each of them, identify actions you can take to accept, avoid, mitigate, transfer or share the risk.

Apply the risk register to actions you are considering (for example, moving house, job or relationship). What is the outcome you really want? Write down the benefits and potential pitfalls over a decade. What are the obstacles you face, and what would it actually take to overcome each of them? What information would you need to make a more scientific appraisal of the risk? What can you learn from others who have taken a similar risk? Can you break down the challenge into smaller ones?

While living in the UAE, I gave leadership training to astronauts Hazza Al Mansouri and Sultan Al Neyadi, the first Arabs in space. I had expected them to explain courage in terms of the intense physical and psychological training they had undergone for their space mission, the isolation and danger, buoyancy exercises, or the extreme G-forces. Instead, they spoke about the agility they needed to learn a new language in order to join the International Space Station. Hazza told me that they 'went up to eight Gs [gravitational force] which is up to eight times our body weight. But it was learning Russian that most challenged us, and most changed us. It was further outside our comfort zone than the astronaut training. But ultimately it made us look at everything differently.'

How? 'I became more inquisitive,' said Hazza. 'I realised that science and knowledge are irreplaceable for achieving progress and prosperity. I learned the skill of coexistence and adapting to live in any environment. If there is one skill that I would like everyone to take away from my experience, then it would be to lose the fear of failure. One who does not fail will not know the path to success. There were so many moments when we felt we were off target or failed physical assessments. So I am not afraid of failure. It was a driving force for the success of our mission.'

During my time as ambassador to Lebanon, whenever I was at a low moment in Beirut, overwhelmed by the scale of the political challenge or agonising over how to keep my team safe, General Sir Graeme Lamb would appear. Normally with a simple, mischievous but useful message: 'Ambassador, proceed until apprehended.'

What does 'proceed until apprehended' mean to a general who has put down his weapons? 'Life is all about failing and succeeding,' he says. 'I recommend a stoic approach to life: I accept what I cannot change. Anyone who has lived a full life and really pushed the boundaries will make their big mistakes. I appreciate fresh linen sheets, as I've slept on rocks. The more you can push those boundaries wide and really reflect on them. Everybody wants to be someone else. I am content with who I am.'

The former head of the SAS is among the most decorated living soldiers. The stories that can be told about him would fill this book and those that can't would fill several more. When I asked what initiative meant to him, he gave an answer you might expect. 'In my line of work it is what it takes to chase a man down. When it all goes wrong the quiet person is the person who jumps up and savages the enemy. When the bullets are flying, the options are fight, freeze or flee. If you wait for guidance or someone to tell you what to do, you really shouldn't be doing what you've been asked to do.'

But when I asked Graeme where he had learned that courage, his answer surprised me: 'Dyslexia was an absolute gift,' he says. 'I was an awful student: I got 3 per cent in my first French exam, and bust a gut trying to improve. I got 1.5 per cent in my second year! In the 1960s my dyslexia meant I was classified as just being dumb. I'd end up alone a lot, reflective. I learned not to get lost, to survive outdoors. At Staff College they saw something and offered me the shot at an armed brigade in Germany. I said, "Forget it, I won't enjoy the job, it's not me, I can't write." I was old enough and ugly enough to know that I think differently because of my dyslexia. What I have which defines me in many ways is the difference in how I look at a problem.' Graeme turned down the dream promotion and took a road less travelled. In the end it led him to jobs and experiences beyond the imagination of most of us.

For artificial intelligence expert Tabitha Goldstaub this willingness to start again, to learn something new, to build on failure, to realign life with purpose, will be one of the skills that sets apart the most employable in the future economy. When I asked her what we needed to learn, she said that adaptability should be high on the list. 'It's no longer the actual skills but the ability to learn new skills. So where a job spec used to have a bullet point list of skills, now we need to have people who are able to learn and adapt to change.' And the secret to her success? Like General Graeme, it was dyslexia. 'It got me to where I am today. Because I'm awful at retaining knowledge and really bad at basic maths, so what I would end up having to do is relearn each time I needed to do a new challenge. It became good scar tissue. And that's pushed me to be comfortable with uncomfortable situations. And that makes it an enjoyable thing learning new skills, rather than being daunting. Nothing ever came easy so I wouldn't get comfortable. I was looking for what was next.'

After his record-breaking crossing of the Indian Ocean in a rowing boat, I asked an exhausted Maxime Chaya why he continued to drive himself to extremes that the rest of us believed impossible. He paused for a moment and said simply, 'Boats were not built to stay in harbours.'

Think again about that eulogy. You will want it all to mean something. What will it mean? What can you do with your time that is truly important? What are you willing to struggle and sacrifice for? What, when you are doing it, makes you forget everything else? What makes a difference?

Only you can answer that.

Once you do, write it down.

And don't leave it unsung.

4

How to Find Your Voice

In the end, we will remember not the words of our enemies, but the silence of our friends.
Martin Luther King, Jr.

Over the years that I worked in 10 Downing Street, I collected a book of advice from world leaders and others, to give to my son Charlie on his fourteenth birthday in 2020.

As the project gained momentum, those I asked for advice would take more time over it. Bill Clinton wrote his out in draft and copied it into the book in beautiful and careful handwriting. George W. Bush took the book away with him, and returned it with his entry and a nice note about how much he had enjoyed reading what others had written. Mikhail Gorbachev got quite emotional composing his entry. Barack Obama commented that Charlie would either be very rich or very clever, depending on whether he sold the book or read it.

The advice they give varies from the idealistic to the practical, and in many cases reveals much about how those who wrote it viewed their own leadership style at those critical points in history. Leaders told Charlie to dream big, to give, and to get to know different kinds of people. Bush counselled against sacrificing his soul for public approval. Gorbachev wrote poignantly about the fact that he might not be around by the time Charlie read his advice.

There were also plenty of entries from sportspeople, authors and celebrities. Carla Bruni told him to play with his dad every day: 'car il à besoin de jouer, même s'il est un adulte'. Footballers David Beckham and Pelé and Olympians Steve Redgrave and Chris Hoy focused on the dogged persistence and hard work that had brought them such extraordinary success. Archbishop Desmond Tutu, Bill Clinton and the Dalai Lama urged Charlie to give.

Above all, the advice was optimistic. The leaders who wrote in the book were genuinely excited about the world that Charlie and his generation would inherit. But the entry that means most to me is the one from my own grandfather, shortly before he died. He encouraged Charlie to find his voice.

For some, the ability to find your voice will be *the* twenty-first-century superpower. Such as the author, poet and former Iraqi refugee Ahmed Badr. 'We need to encourage the ability to turn personal stories into transformative tools that thoughtfully activate and amplify the voices of those living within marginalised communities,' he told me. 'This will be a new kind of engaged storytelling that is community-centred and community-informed, a storytelling that continuously engages and reimagines meanings of agency, and asks: what does it mean to tell our stories on our own terms?'

For Ahmed, 'the twenty-first century is and will continue to be defined by storytellers, for better and for worse. We must be aware of the stories that divide as well as the stories that unite, and critically examine the implications of the stories we hear and see, and those that are made invisible by competing interests. Ultimately, we must be able to create spaces for other stories to exist beside and beyond our own.'

We are surrounded by a cacophony of voices. How can we make ours cut through the noise and resonate with people in a way that supports the sense of purpose we have identified? I've spent time with current and future leaders trying to help them answer this question. The conversation normally starts in the wrong place: how do I get noticed? How do I reach as many social media followers as my colleague or rival? The urge to achieve something measurable and quantifiable can overwhelm all else. But this is the content equivalent of trying to win a contest for eating the most Big Macs.

A much better starting point is to ask yourself: why am I communicating? If the answer to that question is aligned with the sense of purpose we looked at in the last chapter, you will see much faster results. If it is connected to the eulogy values you hold, you will reach people you thought unreachable, and connect with extraordinary allies you would never otherwise have discovered, people who will run through walls for you.

This is why it is vital to know why you are writing or communicating. To entertain? To make the reader or listener smarter, or better informed? Or are you doing it as part of a bigger idea or movement?

Next, make your content authentic, whether you're writing an article, posting a tweet or giving a speech. People need to hear your voice, to experience the human behind the handle. This is why we are so much more likely to follow an individual's personal social media account than their official account. We crave the authenticity.

Then make the content engaging enough to catch – and, crucially, hold – people's attention. To make them lean

forward. My advice to leaders and students is to be a simpli-
fier not a complicator. Try not to use long words, ambiguity,
jargon or fluffy platitudes. Tell simple, unexpected, concrete,
credible, emotional, inspiring stories. Not 'We created a
hundred jobs' but 'Let me tell you about Ben and the work he
is now doing'. Not 'We are concerned about the Syrian
government's actions' but 'I am angry that the Syrian regime
continues to kill its own people, and ashamed that we can't do
more to stop them.'

Mostly, an imperfect message that gets heard is better than
a perfect one that doesn't. It helps to let the ideas drive the
rhetoric, not the reverse. And it helps when we keep it short,
straightforward, purposeful and kind. Or as Ted Sorensen,
John F. Kennedy's speechwriter, put it: levity, brevity, clarity
and charity. It also helps to really understand the people you
are trying to reach. What do you want them to take away?
Whatever the medium, communicate as if you were talking to
them.

As with all the survival skills I've described so far, a major
step towards creating great content is to observe, learn, practise, teach. When you see great writing, speeches or
communication, try to take it apart to see why it worked.
Study those lessons and practise them in the way you would a
musical instrument. This can even apply to social media. I
asked Instagrammer Jeremy Jauncey what had been the key to
his success in creating content that excited and inspired
people.

'The first most important thing is to be able to explain yourself,' he says. 'What is your story? Because if you can't explain
your story in a confident way, why would anybody else follow
you, vote for you, hire you, invest in you? So that is a skill
that – it doesn't matter who you are, you have to absolutely
master. Second, an ear. I think it is being able to form a cohe-

sive thought, articulate it, and then listen to someone with a different perspective and change your thought process or be able to engage in a dialogue. The third is to find your cause. Young people are cause driven, and respond to communication that speaks to that. They genuinely care about things that matter to the planet. They're living it day in day out and they believe they can change it.'

So think about the call to action. What do you want to happen when you've finished? In Gordon Brown's favourite (though historically suspect) story, Cicero was said to be a great speaker because he left his audience in awe of his rhetoric and intelligence. But Demosthenes was a greater speaker because he left people thinking, Let's march.

Those of us who heard this story frequently had no idea if it was true, but it had the effect he wanted. The best content is action not reportage, purpose not platitudes. It is about changing the world, not just describing how it looks.

I used to hate making speeches. I was terrified, even way into my thirties when I was doing them as an ambassador. I now make them for a living. Spoiler alert: I'm still terrified. Fortunately it turns out that I am not the only one. I watched Tony Blair close up making a speech in Nigeria. By that stage in his third term as prime minister, he was easily the most gifted communicator of his generation. The speech was not going to define his career or hit the news. And yet, backstage as he prepared to go out, his hands were shaking. Even Tony Blair gets nervous before his speeches. I remember telling myself then never to worry again about feeling nervous before speaking. I still find that even the most straightforward public speaking requires courage.

Mark Twain's advice should be on the front of every manual on speechmaking: 'Courage is resistance to fear; mastery of fear – not the absence of fear.'

I now teach a class on giving speeches. There are ten pieces of practical advice that I've compiled for my students from great speakers over the years.

1. Put yourself in the shoes of the listener. What question do they want answered? For many, this is 'So what?'
2. Understand the purpose of the speech: why are you really doing this? Never forget that it is as much about *why* and *how* you say it as *what* you say.
3. What is the essence of your argument? What is the central idea, the takeaway/tweetable line? What will people remember in a week? If you can't decide what that is, no one else will hear it.
4. Understand the physics: observe and use your nerves; think about how to enter the room, stand and move. Think about the pacing, pauses and volume. You are almost always going too fast.
5. Know and trust your content. Never under-prepare. It takes effort to look effortless.
6. Begin it well. The first line sets the tone. Are you aiming for humour? Confidence? Praise? Rapport? Reassurance? Thanks? Don't waste that chance by tapping the microphone awkwardly and asking if it is on.
7. Narrative. Tell a story.
8. Be short. Remember that three messages are normally all that you or they can hold in our heads.
9. Feel the love, engage people personally (for example, use their names), justify their attention, respond to the room. For most of the speeches that most of us do, the people listening want it to go well. They're on your side.

10. End it well. What do you want from the last line – applause? Laughter? Fist-clenching camaraderie? Don't waste that opportunity with an excuse or an apology. Find a way to link it back to the opening to give people a sense of a narrative arc, of closure.

As ever, learning this skill starts with observation. Abraham Lincoln's Gettysburg Address is the most famous speech ever made, given at the dedication of the Gettysburg National Cemetery during the American civil war. Why does it work? For me, it succeeds because it is short and simple, less than two minutes long when spoken. It has three parts: looking back to America's founding, considering the challenge of the day, and setting a vision for what can be. It has an attention-grabbing opening: 'Four score and seven years ago our fathers brought forth on this continent, a new nation, conceived in Liberty, and dedicated to the proposition that all men are created equal.'

And it has a memorable close: 'From these honored dead we take increased devotion to that cause for which they gave the last full measure of devotion – that we here highly resolve that these dead shall not have died in vain – that this nation, under God, shall have a new birth of freedom – and that government of the people, by the people, for the people, shall not perish from the earth.'

I use a version of this with those I teach and coach: how to give a speech any time, anywhere. It has got me through countless daunting moments, often when I don't have the luxury of time for preparation. You can deliver a speech any time, anywhere by breaking it into three parts: past, present and future.

We were here.

Now we are here.

This is where we are going.
Try it.

Our family once lived and worked as arrow makers. Imagine their lives, the skills they passed on, the moments of war when their craft was tested. Now we are coming together for Christmas with another year of adventures behind us. In her design class, Sophie is doing learning skills that those arrow makers would have found extraordinary. Through his military service as a peacekeeper, John is helping to contain the spread of weapons in modern conflicts, where once we made them. And imagine what we will build together in the future. Imagine the family reunion in thirty years' time, the new children and grandchildren who are kind and curious and brave. What will they be crafting? What mark will they be making in the world?

Or you can experiment with more mundane subjects.

Think of the humble banana. Imagine how it felt to be the first European to see one, to taste one. And now every country in the world can experience the banana. Look at the recipes it inspires, the nutrition it provides. Unlike that first European, we now understand the importance of the potassium it provides. And we also understand the need to ensure that those who grow and sell bananas are treated ethically, with rights and dignity. And let's think now what the banana can be in the future. A bridge between cultures. A fruit that can inspire new ways of cooking.

You get the idea. Past, present, future. Give it a go for the next terrifying speech and I guarantee that you will be amazed at the results.

Again and again, the individuals I interviewed for this book fell back on another vital way to find our voices: get a mentor. Baroness Cathy Ashton was the European Union's first High Representative for Foreign Affairs and Security Policy and one of the most effective negotiators of the twenty-first century. She learned her skills as a trade union negotiator, and through having to break glass ceilings. Her father advised that if she didn't secure a university place, she would be a shorthand typist. She told my students, 'If I can be a baroness you can be anything. Grab – we call them mentors – but they're really just grown-up friends, people who will tell you "you probably can". These are the people who can help you to be heard. Normally at just the moment when you worry, as we all do all the time, that you probably can't.'

Roger Federer is the most gifted, graceful tennis player of all time. He has a tennis coach. Luciano Pavarotti had a singing coach. Muhammad Ali had a boxing coach. He never stopped learning. I have his words on my wall: 'The fight is won out there on the road, in the gym, long before I dance under those lights.'

All the most successful people in business, sport or politics have a mentor or coach. They don't assume that they are the finished article. As with seeing a therapist, it is often just the act of recognising the need for help that makes the biggest difference. A coach or mentor can help you think like an apprentice. They allow you to step away and reflect on finding your voice. They give structure to your approach to your development. Many of us, if we're honest, still think that our development is something we do between jobs or on the occasional training course. It is not.

I asked coaching and mentoring guru Claudine Menashe-Jones for her tips on this challenge. She said that listening was the most important part of mentoring and being mentored. 'We need to get out of advice-giving mode,' she says. 'As mentor there's a tendency to be thinking of what you're going to say next, how you're going to help, what fix or story or guidance you can offer. The pressure is on to bring insight and wisdom ... and so you're thinking about yourself, not about the person you're supposed to be listening to. You risk not picking up on the nuances in language, expression and tone, and hence on the things that your instinct would tell you it's useful to explore further. The classic example here is the sports coach who instead of shouting "Watch the ball!" says "What's happening to the ball? Which way is it spinning?"'

In practical terms, this means making sure that you find someone who you feel really listens to what you say and asks you questions that truly make you think rather than someone who only tells you what they did or would do if they were in your shoes. As individuals, we don't need to be 'filled up' – with information or answers – by somebody else: we have all the resources and answers we need to develop or to achieve our goals. Clearly this doesn't mean that we can't learn new skills or absorb new knowledge, but it reminds us not to underestimate what we're already capable of, what we already know and what capacity we have for growth.

And we should not assume that mentoring like this is something older people do to younger people, or bosses to underlings. Companies such as Cisco have schemes in which younger people more comfortable with technology coach those older and more senior than them. Don't fall for the assumption that only people more senior than you can help you master or hone new skills. When I was ambassador I wanted to improve my attention to detail – my coach was the embassy's junior accountant.

Again, this need not be complicated. I recommend three steps for finding a mentor or coach. First, think about the skill you want to develop, and why. The process only works if you feel that sense of purpose and agency.

Second, observe people to identify those who do it well. They won't necessarily be conventionally successful role models.

Third, and often the hardest, ask them to help. This takes courage. Most will say yes. They are more likely to agree if you have thought carefully about what you want to learn, and why. Many people pick a coach or mentor not because they think they can learn something from them but because they think that knowing that person better will help their career. In some cases it might. But most would-be mentors can spot this a mile off.

Once you get to that point, there are three vital aspects to the best coaching and mentoring. It is vital to agree expectations. How much time will you both bring to the project, and for how long? Then to set clear objectives for the skill you are working on. And finally to create an environment of complete trust and honesty.

As part of that process, finding our voices will also require us to be better negotiators. Negotiation is an art not a science. But it can be practised. In reality, while most of us won't be winning many Nobel Prizes for our efforts, we negotiate most of the time. As we navigate the challenges ahead, we will find ourselves negotiating over access to the local water supply, whether to open up schools even if it increases the spread of pandemics, or who gets to use which bathroom. Those that can do it most deftly will increase their survival rating. How can we practise?

Good negotiators learn to know their interests. This sounds obvious, but it is striking how often humans arrive at international conferences or negotiations over their contract without a clear sense of the outcome they want. Sometimes they come with unrealistic negotiating positions, or none at all. Often they are unable to rank their demands or expectations. Cathy Ashton, who as the EU's vice president negotiated vital deals in the Balkans and Iran, told me that the best negotiators visualise a box that captures their upper- and lower-end expectations. Finding which parts of that box overlap with your interlocutor's box is the sweet spot of the negotiation. The effective negotiator will have real red lines, and artificial ones. There will be cards that can be conceded. Before any serious negotiation, it is therefore worth building in 'negotiating fat'. The successful negotiator makes concessions. In the case of the Versailles Treaty in the aftermath of the First World War, failure to do so contributed to the Second World War. Good negotiators sometimes even let their opponents win. Ex-diplomat John Ure suggests that 'the best diplomatic victories are those when everyone goes away thinking they have won. Diplomacy is the art of building ladders for other people to climb down.' Or as the American writer Sue Monk Kidd put it, 'If you need something from somebody, always give that person a way to hand it to you.'

Identifying a common vision for what you both want also helps. When this vision exists, it is easier to manage those moments of fragility when a terrorist attacks or your partner fails to turn on the dishwasher. If the British government had demanded disarmament of the IRA before peace talks could start, we might still be at war with them today. If we had not demanded President Assad's departure before talks could begin on Syria, the country might not be in pieces today. The negotiation over which greenhouse gas emissions to cut is easier when we remind ourselves that we share a common interest in reducing climate change.

The negotiations on the big identity questions ahead of us also require us to leave our echo chambers. Otherwise why do it? Too often, especially on social media, engagement with opponents is presented by critics as some kind of reward for those opponents. Sometimes, when the moment is right, we should instead take Admiral Nelson's advice: 'Never mind manoeuvres, go straight at 'em.' As the philanthropist William Sieghart, who has been involved with many painstaking peace-making initiatives, told me, 'Enemies ultimately have to engage in order to end conflicts. It's astonishing how many people once demonised can have their horns and tail removed at a later date.' We should remember this more in the increasingly confrontational identity arguments that consume much of social media.

Compromise has become a dirty word, associated with spinelessness. Heaven forbid that a leader should be presented in the media as being like Neville Chamberlain, the British prime minister between the wars and architect of appeasement. Yet short of military solutions, which we're less keen on now, compromise is often the only way to make progress. It need not mean weakness. 'Tolerance implies no lack of commitment to one's own beliefs,' said JFK. 'Rather it condemns the oppression or persecution of others.'

Finding our voice takes time. It is a process, not an event. Former diplomat Sir Simon Gass, who led most of the Iran nuclear negotiations for the UK, says that effective negotiators conserve their energy when it is not needed, marking time or letting others fill the space until the context changes. Much of the Iran negotiations felt like this. As the Iranians worked out, deadlines are usually more flexible than they appear. Veteran peacemaker and former US senator George Mitchell described one such lengthy negotiation as 'Seven hundred days of failure and one day of success'.

Finding your voice is not easy. As part of our NYU research we asked a group of retired political and business leaders to tell us how they did it. They identified ten pieces of advice they wished they had been given when they were twenty-one years old. We challenged them not to pass on classic platitudes, but to be brutally honest. After much argument and debate, here is what they concluded.

1. People-watch intently and indiscriminately – and remember you learn as much from observing people doing the wrong things as from the role models.
2. Find a way to feel confident (not the same as showing confidence – if it is not genuine then you will quickly get found out).
3. Feedback is a gift. Welcome it and accept it gratefully even if it makes you a bit uncomfortable at the time.
4. Identify your unique selling point, something that distinguishes you from others. In a competitive environment, it helps to have something that is a bit different. Find what you do for people that no one else does.
5. You cannot change your underlying character. Don't try to, or you will come across as fake. But be ready to flex your style to meet different situations. Be you. Compromise on the things that don't really matter to you but not on that.

Hitch your wagon: identify a good and influential boss or senior figure; and make yourself indispensable to them.

7. Impostor syndrome is normal – anyone who doesn't feel it is probably stupid or arrogant.

8. Do the things you enjoy. And enjoy the things you do. Not just does that make the work easier, but it makes you an enthusiast. People remember that.

9. Seek out the person in the room who everyone else regards as the least important.

10. Find what it is you want to say and stand for. Something you really believe. Don't be afraid to articulate it and fight for it. You may not win, but you may well outlast those who beat you and what goes around comes around. You don't want to reach your fifties and realise that all you have ever done is transacted business. You will want it to mean something.

You can see this spirit in the leader of a new era of climate activism. In finding her voice, Greta Thunberg has mobilised hundreds of thousands of people across the globe against climate change. She has been named *Time*'s Person of the Year and was nominated for the Nobel Peace Prize in 2019. She has shown herself willing to speak truth to power, sparring with world leaders such as Donald Trump, who publicly mocked her Asperger's syndrome.

Greta's voice is unusual but not unique. Where does it come from? 'When you are actually bravest, you don't necessarily realise it,' she has said. 'You are just doing what you know you should be doing. You are where you are meant to be. It is only courageous later when you look back and think, Oh wow. When you know what you believe, the bravery is some-times just there.'

In *Why We Kneel, How We Rise*, West Indies fast bowler Michael Holding recounts how he suffered 'barbaric' abuse as a cricketer in England in the 1970s and 1980s. His answer to that problem is better education, to help people understand the causes of racism and the factors that sustain it. 'Until we learn that and we understand that,' he told the BBC in 2021, 'we will not be able to tackle it and get rid of it. We need to re-educate people so that they learn the true history of mankind.'

Holding's powerful piece to camera on Sky Sports following the murder of George Floyd in police custody in Minneapolis was an extraordinary example of a public figure finding their voice. Asked about it afterwards, he described how the arguments had built up over decades: 'When the moment came, I did not think about what I needed to say. I just felt a powerful conviction that I needed to say something. And it just came tumbling out.'

When you find your voice, you will want it to mean something.

When the way we communicate is aligned with our values and sense of purpose, it is no longer just about building visibility and legitimacy. It becomes activism. And activism requires some allies.

5

How to Find, Grow and Mobilise Your Community

A tribe is a group of people connected to one another, connected to a leader, and connected to an idea. For millions of years, human beings have been part of one tribe or another. A group needs only two things to be a tribe: a shared interest and a way to communicate.

Seth Godin

We are hardwired to be tribal, whether we like it or not. Traditionally these tribes were about shared genes: we had a collective survival instinct with our fellow clan members. For much of the twentieth century they were more likely to be formed by school, university, profession, class or location.

The new tribes are more fluid, more dynamic and less predictable. But they will shape our survival as much as the old ones did. This chapter looks at how we find and form these new tribes, how we find our voice and build legitimacy and influence within them, and how we can turn that into real, lasting change.

The last decade has seen us think much more about our tribal identity. Issues such as the Iraq war, Scottish independence, Brexit or Donald Trump seemed to demand a more binary approach. Which side are you on? Are you with us or against us? But the new communities are developing in other, more subtle ways. Just as our ancestors sought to create groups away from the chaos of the public square, this desire to gather elsewhere will accelerate as people lose confidence and trust in the public square of social media and political debate.

A quick way to assess which communities you are currently a member of is to look at your WhatsApp groups. My most-used groups are immediate family; extended family; the Strollers cricket team (a triumph of hope over expectation) that rarely now plays cricket; a group of school friends that used to go on an annual barge holiday; and the music-themed group of university friends in which we spend too much time creating Spotify playlists. Even before the lockdowns, I had far more social interaction with those groups than with any offline community beyond my wife and children.

After these immediate communities of family and banter, I have more professional groups on education, diplomacy, the Middle East, technology. Increasingly you find that these tribes convene around an email distribution list or podcast. As we navigate new social interactions we are constantly scanning the horizon for these points of connection, allegiance and affiliation: a fellow despairing West Ham fan; a devotee of the third series of *The Wire*; the colleague who knows all the catchphrases from the *Tailenders* podcast; or another exhausted parent on the school drop-off.

These realities of identity are of course much more complicated than they often appear, or than the political tribes we are invited to be identified with. And the example of sport gives us a clue as to the pitfalls of defining our worldview on the basis of national identity. As the brilliant former Irish ambassador to the UK Bobby McDonagh has warned, when comparing the way we develop loyalty to sports teams or nations, 'I have learned that the emotional allegiance that flows from a sporting identity clouds my judgement, both as regards refereeing decisions and predicting outcomes.'

One of the most exciting areas in which new communities are developing is one that many of us might find impenetrable: TikTok. Most people over thirty dismiss the memes and hashtags of this video-sharing app as bizarre and childish. But cultural ethnographers are fascinated by the way in which people are now finding their communities and organising themselves on it. These dynamics may even be shifting the topography of how we organise society. Elena Liber, co-founder of a Goldsmiths, University of London project on TikTok, describes how she became interested in the app during the first lockdown in March 2020. 'Friends began sending me videos of people dancing, cooking food, growing houseplants and telling jokes,' she says. 'I was fascinated by how so much creativity could be captured by a one-minute video. Following the murder of George Floyd and the subsequent Black Lives Matter protests, I was amazed at how protest and resistance was organised online, especially on TikTok and how TikTok became a space where experiences could be shared, educational material disseminated and protest could be organised. Similarly the women's rights protests in Poland and the pro-democracy protests in Belarus have a significant online presence.'

So a good starting point to understanding the new ways that we are forming groups – whether or not of protest and resistance – is to observe your current communities and assess whether these are indeed the tribes you want to spend your time in. And if not, to think about how to convene, build and mobilise groups around the issues that are most critical to your sense of purpose. If your issue is the crossover between Jethro Tull music and political reform, how do you become the convener for those discussions? Who do you want to recruit to that group and what are the incentives for them to join?

One aspect of this effort to find a voice and a community that has been overlooked is how much harder it is for minorities. It might be that others mistakenly pigeonhole you. Or that your different perspectives are less likely to be heard. So no curriculum or education is complete that does not help young people find a way to make space for difference. The national and international debates stimulated by Black Lives Matter and the Me Too movement have helped. But we have much further to go. Too often we are left suspended between knowing we need to get better at inclusion and the fear of getting it wrong. We'll come back to this vital debate – and essential survival skill – in chapters on coexistence and being a good ancestor.

Writer Simon Sinek explained to me what happens when we succeed in finding that tribe: 'There's a recurring pattern – we get together amazing talented people, they go "Holy shit, I've found my tribe, my people" and there's this moment of absolute exhilaration. The puzzle pieces come together and we create a picture we didn't know was even possible.' We feel heard, we feel less unusual. It is energising.

But Simon also described the danger of what often comes next. Soon after the initial excitement wears off, there are negotiations and trade-offs around money, power, time, influence, platform. 'Then that moment of initial bonding fractures again … and there's a temptation to just take my ball and go home.' So we have to develop the skill of not just finding and building a community but building the trust and sense of shared action to sustain it.

Introductions are the key to building a community. Traditionally, a personal introduction was one of the best ways to help someone. The internet hasn't changed that. But just as an offline introduction could vary in quality, from a brief handshake to a dinner, the quality of online introductions can differ hugely. Compare the email you get saying 'I wanted to connect you and Dave' to the one that gives a paragraph on you both, adds a line on why the emailer wanted to connect you, and offers to help in any way in kick-starting a meaningful relationship. These introductions are rare and valuable. They are worth making time for – to ask for them and to offer them.

Arriving as a migrant in Abu Dhabi, I built a professional portfolio on three introductions like that and one existing professional friendship. In each of the cases, I would now drop everything for the person who made the introduction. And while it is right to recognise their value, these kinds of exchanges are better for not being transactional. But they can create a powerful bond. Building networks and connecting with kindred spirits in this way is an art and a science. As with other survival skills, you can look to observe those who do it effectively, and practise what you see. We often notice more of the art: what looks like effortless confidence or a genuine interest in other people.

Here are three ways I've observed for improving the science of connecting people. Keep the introductions and check back in with them to see what happened. Take time to really think about why the connection matters for both of them. And when you have a breakthrough or success as a result of an introduction, remember to go back and thank the person who made it happen. Building a tribe takes patience, trust and humility.

So we have found our cause, started to build legitimacy and alliances, and developed the craft of getting our words read and our voices heard. Once you are more confident that you are in the right communities, the next point is to mobilise that community, to work out what actions you can take together.

Finding our inner activist need not mean delivering the stirring speech or manning the barricade. Communications coach Sho Konno helps people and organisations improve their activism. 'Until relatively recently, you would often hear people say "Well, I'm obviously not an activist, but …". Now half the brands in your supermarket proudly claim to be activist,' he told me. 'They're both wrong in the sense that it is not a formal position you can earn and then award yourself the title of "activist". If you have signed a petition, stopped buying something for an ethical reason, or supported someone's protest online, you have probably already taken part in activism.'

Sho's tip is to think about activism by asking questions like 'Why is our world the way it is?', 'How could our world be better?' and 'What can we do to make that change for the better?' That last question – putting the 'act' in 'activism' – is crucial. The first step in activism is simply finding other people or groups who care about an issue you care about, and asking them how you can join in. If that group doesn't exist yet, start it yourself.

Kirsty McNeill, a former colleague of mine in Number 10, now runs Save the Children's campaigns. She believes that the pandemic made more of us activists, from Marcus Rashford to those organising rotas to deliver food or to check on older neighbours. 'Part of what is going on here is the public's sophisticated understanding of the coronavirus – that the experience might be universal, but it is not uniform,' she says. 'We understand that there are people in precarious employment in every country, parents struggling to put food on the table in every country, children trapped on the wrong side of the digital divide in every country. We need to turn the new neighbourliness into the new normal, while helping people draw connections between their new local involvement and the need for active citizenship at a national and global level.'

Polls suggest we feel that we are indeed likely to come out of this crisis more connected and kinder than we went into it, but this effect is much more pronounced for people with whom we have direct social contact. 'The more we know people, the more we trust them', Kirsty argues, 'and the street or estate where we live is now full of people we newly know.' Ten million citizens have chosen to spend at least three hours a week caring for one another.

Covid-19 reminded us that there are activists among us everywhere, from the nurses, cleaners and teachers heading towards the virus while we stayed home to the scientists working on the vaccines that allowed us out again. The pandemic created what Professor of Globalisation and Development Ian Goldin calls 'an unprecedented resonance in the human experience … for the first time, the world is sharing the same experience at the same time'.

But the crisis also showed that we can do more to stockpile compassion for those individuals, communities and nations that the virus hit hardest. We must do much more to sanitise our political debate of the xenophobia and selfishness that rose in some quarters, and that movements such as Black Lives Matter exposed. We can keep calm and carry on with the quiet, patient work of educating the next generation, including the Alexander Flemings who will discover the vaccines of the future. We can choose not to wash our hands of the long list of other challenges we need to confront. Human development has always depended on the energy of small groups of people, often radicals derided in their own time.

Can this be learned, or taught? Can we hone this skill of activism throughout life?

The greatest leaders can do three things in the service of a cause: inspire others with a vision, bring people with them, and then put in place the plans to deliver on the vision. Most of us can't consistently do all three. I think the prime ministers I saw close up – Tony Blair, Gordon Brown and David Cameron – could each do a different two of the three. As I'm still a discreet ex-civil servant, I'll have to let you figure out which one each was missing.

Mobilising our community also requires us to be clear on the vision, to bring people with us, and to get the plans in place. This might sound daunting, but many of us are activists already, usually without realising it. And the lockdowns might have helped both to build stronger communities and to spur us on to become more effective activists.

What does that look like in action? It looks like the public campaigns that brought down Augusto Pinochet's murderous regime in Chile in 1988, the authoritarian regime in Czechoslovakia in 1989 or Muammar Gaddafi's dictatorship in Libya in 2011. But it doesn't have to be as dramatic or as violent as facing down a dictator. It looks like the campaign of Californian estate agent Candy Lightner to make drink driving illegal, after her thirteen-year-old daughter was killed in a drink-driving accident. It looks like Eddie Mabo's successful 1992 campaign to gain recognition for Aboriginal land rights in Australia. Or the peaceful protests of the Liberian women who brought a brutal thirteen-year war to a close.

Some of us are humanitarians: if we see a bully beating someone up at school we go to help the injured person. Others are activists: they confront the bully. We need more people to do both. Acting alone is difficult. Acting with others reduces the risk.

George Mpanga is better known as George the Poet. In 2019, his podcast – mixing music, storytelling and poetry – won the Peabody Award and was named Podcast of the Year. Of Ugandan origin, George studied sociology at Cambridge University before becoming an influential activist. He told me that the effort started from a recognition that young people were serious about change but not always sure how to make it happen. 'We need to help them use their antennae, help them find their voice, find their conversation, make an impact,' he says.

For George, part of the activism has been channelled towards changing education. This was important to him because he had seen the way in which the education system could subtly reinforce and perpetuate privilege. 'We're getting the diagnosis wrong because we fail to take account of cultural history, people's socioeconomic context. You can't just throw money at the problems we face. You have to find a way to listen.'

Malala Yousafzai loved school. But everything changed when the Taliban took control of her town in Swat Valley and said girls could no longer be educated. 'I spoke out publicly on behalf of girls and our right to learn,' she said. 'And this made me a target. In October 2012, on my way home from school, a masked gunman boarded my school bus and asked, "Who is Malala?" He shot me on the left side of my head.' After months of surgery in the UK, Malala recovered and has devoted her life to getting every girl twelve years of education. In 2014, she became the youngest-ever Nobel laureate.

I asked Malala what activism means to her. She said she saw it every day in campaigners for girls' education. 'I was just the one on that bus that day. There are millions of us quietly fighting for our education, and taking the same risk. But the courage we share is that we know it is worth it. And that even when some of us fall, more will take our place. Courage is easier when you believe so strongly in a cause.'

Jamira Burley is another example of how younger activists are making their desire for change part of their day job rather than a protest to attend at the weekend, a hashtag to share, or an organisation to join. Jamira calls herself a next-generation community social impact activist. Driven by the murder of her brother Andre, she organised an anti-violence programme at her high school in Philadelphia, which reduced the rate of violence by 30 per cent. In public recognition of her efforts, she received a $50,000 grant from the state governor to implement the programme in the ten most persistently dangerous high schools in the city. Jamira went on to be the first of her fifteen siblings to get to university. 'You are faced with a choice at that point,' she told me. 'Do you pull up the ladder or do you spread the opportunity, and use your position to tackle the underlying causes of injustice and inequality.'

We'll need a similar approach to the culture wars that otherwise threaten to polarise society and hold us back in our search for community and activism. In later chapters we will look at the systemic injustices and inequalities we must take on if we are to become better ancestors. Yet, as a recent Fabian Society paper explored, on current trends, it seems 'much more likely that we will see progressive movements and political parties distracted, divided, demoralised and defeated by those pursuing a strategy around the so-called "culture wars".'

The paper's authors, Kirsty McNeill and Roger Harding, define such issues as 'those concerned with identity, values and culture which are vulnerable to being weaponised by those concerned with engaging and enraging people on an emotional level, not coming to a policy settlement on a political level'. Recent examples include whipped-up controversies over whether an Oxford student common room should have a picture of the Queen, or whether 'Rule Britannia' should be sung at the Last Night of the Proms. Culture wars are often encouraged by those with a genuine grievance or sense of inequity; to get more visibility and attention; or just for the pleasure of trolling – seeing others in pain.

Why do these culture wars matter to our effort to live together in the future, to survive? For McNeill and Harding, they risk becoming a distraction from the actual issues that we should be addressing. They set us against each other and demoralise us. And they risk making the public square toxic, especially for people from marginalised groups. In response many of us might choose to ignore the culture war arguments in the hope that they will simply fizzle out. Or we might pursue a more adversarial approach. But neither seems to work.

How, then, can we respond? McNeill told me the answer would be based on four pillars. First, constructing a more confident, inclusive vision of the future that more people can see themselves in. Second, 'renewing our democracy by selecting and rewarding leaders who exercise restraint, value pluralism, defend institutions and norms and are willing to regulate technology'. Third, being ready to call out those peddling the culture wars in order to distract or divide. And, finally, to build movements 'that are inclusive in composition and culture so that everyone, especially people who have been marginalised, has a stake'.

What does this look like in practice? We can see it in the response to campaigns and individuals like footballer Marcus Rashford and Captain Sir Tom Moore. Moore became the talisman for support to the UK's health service during the pandemic. Rashford rallied a coalition from across communities, regions and socioeconomic groups to tackle hunger in schools. He set the agenda for the UK government, at one point in 2021 having to post on social media that he could not fix a policy problem until after his match had finished. Rashford has captured the way that activism starts with self-awareness: 'Whether it be the colour of my skin, where I grew up or how I decide to spend my time off the pitch, I can take critique of my performance all day long ... but I will never apologise for who I am and where I came from. I'm Marcus Rashford, a twenty-three-year-old black man from Withington and Wythenshawe, South Manchester. If I have nothing else, I have that.'

Rashford is not the first sportsperson to have become a model for activism. In front of Adolf Hitler at the 1936 Berlin Olympics, the US athlete Jesse Owens won four gold medals and challenged his host's myth of genetic supremacy. Muhammad Ali, the BBC's sportsman of the twentieth century, was a sublime boxer, but he also blazed a trail for equality and justice. The US sports journalist William Rhoden has described how 'Ali's actions changed my standard of what constituted an athlete's greatness. What were you doing for the liberation of your people? What were you doing to help your country live up to the covenant of its founding principles?' By taking the knee before an NFL game in 2016, the American footballer Colin Kaepernick showed how relevant this idea was. US soccer player Megan Rapinoe's activism on LGBTQ equality has since reminded us of the way sportspeople can successfully confront prejudice.

Activism doesn't just come from the world of art, civil society, sport and NGOs. Mark Carney, the former Governor of the Bank of England, is proving an unlikely but powerful activist. His time in the top echelons of the financial world convinced him that degraded institutions and moral flaws in the prevailing economic model were ultimately responsible for global financial instability, the Covid-19 pandemic and the climate crisis. So he left that world to become a UN climate envoy and to press for a model that prioritises a different set of values, including solidarity, fairness, responsibility, resilience, sustainability, dynamism, humility and compassion. He argues that in any decent society there are three components to fairness that need urgent action: that between the generations, that of income distribution, and that of opportunity. Essentially, his activism aims to bring humanity into our economy and society, starting with encouraging companies to put purpose at the heart of their business models.

But as Carney recognises, this kind of change won't succeed if it is top down. I asked Jeremy Heimans and Henry Timms, authors of *New Power*, how we can pull together the right groups and messages to deliver meaningful change. They showed me that the best ideas spread sideways. Power used to be more closed, inaccessible, top down and exercised carefully. Think of a traditional financial currency. Traditional power values were more formal and managerial. Power thrived on competition, confidentiality and exclusivity.

For Jeremy and Henry, new forms of power are more likely to be open, participatory, driven by individuals. Change can therefore happen in new ways – you no longer simply have to convince the people at the top. Networks can be faster at spreading power than hierarchies. And to get more people involved, we can be influenced by the way that people increasingly place a higher value on experiences they can shape. That works best in conditions of openness and transparency. Think of Airbnb or the Ice Bucket Challenge as models for encouraging engagement and activism. New forms of power can increase when you give power away. Women are much better at this and make better coalition builders; less macho and more Merkel. Barack Obama's more inclusive and empowering 'Yes, we can' would be the motto of the new power movement, rather than the autocrat's 'Only I can do this'.

The heroes and villains of the new power landscape are not always obvious. For example, Facebook and Uber were built to a huge scale by extraordinary networks of connected individuals. Yet they have tended to default to old power values when they come under pressure. They become more secretive, more top-heavy, more controlling, more intimidating. Meanwhile, the US's National Rifle Association has managed to harness new power to preserve its corrosive influence on America's politics and playgrounds. It has built a sense of a collective, with a low entry bar that allows its supporters to feel they are part of something bigger and more powerful. It makes them feel that they are part of a tribe.

The global reaction to George Floyd's murder under the knee of a Minneapolis police officer in May 2020 demonstrated the way that activism is evolving. It led to long overdue tangible change in the oversight of police departments. The impact was partly the result of collective shock and outrage. But it was also because the Black Lives Matter movement had built a coalition, and clear principles and objectives, over almost a decade. As a result, BLM gained unstoppable momentum. This is a new phase for digital activism. Rather than see social media hits as the point of the campaign, successful movements hold fast to the idea of social media as just the vehicle for the message. The campaign itself still needs to be planned, built and organised.

How do we build our own campaigns around the issues that matter to us? We need a strong central message, an honest appraisal of the strengths and weaknesses of our position, and an understanding of which tools will help us reach our potential allies and engage our opponents. We need to embrace the anarchy, assuming that plans will change, adapt. Like the Live 8 movement that built support for significant anti-poverty funding in the run-up to the 2005 G8 summit, the most successful campaigns create cliff-hangers and moments of jeopardy, which are more likely to get the attention of those we need to put pressure on their leaders. There is an element of storytelling to them.

There is much we can take from these models as we build the societies, governments, movements and businesses of the future. In the slow and often frustrating work of trying to build a happier and better society, the key will be to see individual citizens as a vital part of the effort, not just as beneficiaries of government or outsiders. We should trust more in the wisdom of crowds.

Used wisely, social media can be a huge asset to purposeful activism. We need to spend as much time taking and holding the online public square as the Tahrir or Tiananmen Square you can find on a map. Demonstrators in Hong Kong exchange protest guides on how to avoid facial recognition. Campaigns from the 2011 Arab Spring uprisings and Occupy to France's Gilets Jaunes and Black Lives Matter draw strength and street craft from each other. This might have happened with activism in the past, such as over the Vietnam war, but now it takes place at pace and scale. Today's movements don't have to build the same structures in order to mobilise. This gives them great agility, and the ability to act like insurgents, surprising opponents. The challenge, as we saw in Libya and Egypt after their uprisings swept dictatorships aside, is how to fill the vacuum that they leave behind.

If we apply the ideas on turning our communities into places of activism, we have a better chance of surviving the challenges this book has described. We're not done yet as a species. But we are also going to need to spend more time learning how to live together.

6

How to Coexist

For eons, humans have struggled to discover less destructive ways of living together.

Margaret Wheatley

I worked for four years as the British prime minister's adviser on Northern Ireland. I was with Gordon Brown through weeks of painstaking, exhausting and often infuriating negotiations between Republicans and Unionists. At moments it felt like they united only when they saw a collective opportunity to get one over the UK government. And I was with David Cameron as he put together his extraordinary, and very personal, response to the Saville Inquiry's report on Bloody Sunday. During these moments, and my time spent as ambassador to Lebanon, I learned the most about the science and the art of reducing violence and conflict, about the mechanics of living together.

In Northern Ireland it was in the end not the politicians, churchmen or men of violence who really made the peace, but ordinary men and women. In Lebanon, at one reconciliation event, I was introduced to two people who were promoting direct links between individuals across the divide in places that had been cut deep by conflict, such as Rwanda, the Balkans and South Africa. The room was full of trauma, grief and pain. Assuming they were a couple, I asked why they had come.

'My father was killed in a terrorist attack,' said the woman.

I sympathised, feebly, one of those moments where you feel the inadequacy of words. 'And you, sir?'

'I was the bomber.'

I think of this moment whenever I see arguments becoming polarised on social media, or anger spilling into the streets. Much of Donald Trump's agenda, from the rhetoric to the wall and the separation of immigrant families stood in stark juxtaposition to the courage shown by the example of a grieving daughter finding a way to work with the man who had killed her father. Indeed, perhaps the most telling moment of the first Trump presidential campaign had come when he directly attacked the grieving parents of a US soldier killed while fighting for his country.

Perhaps the greatest danger is posed not by the nuclear bomb, environmental catastrophe, the superbug, the robot age or the crazed terrorist, frightening as they all are. The greatest danger may in fact be the loss of our desire to live together. As we face a period of migration, flux and unpredictability, what then are the key practical lessons we can draw from politics, philosophy and religion on how to coexist? This chapter looks at how we can learn more from our shared history; at the moments when our ability to live together progressed; at the battles ahead, from how we design our communities to how we negotiate between old and young. It looks at practical ways to develop the courage we will need.

We have to change the way we teach and learn our history. During the two years that my NYU research team spent asking young people and parents what was missing from their education, a sense of global history, ethics and civics came up in every single group. One US parent lamented that 'at the moment they ONLY study American issues and wars and presidents of the USA. We spend time teaching her about other countries and how they are run and politics, crises in other countries. These kids have no idea where anything is on a map outside the USA.'

Young people told us that there was a growing need to understand their country's role in a global historical context. They wanted to study tolerance and openness towards different groups, to understand how to manage disagreement. 'In an increasingly polarised world, where filter bubbles separate us from those we disagree with,' said New York University student Atoka Jo, 'listening and learning from others who hold different opinions will become even more important if we want to build consensus and cohesion in our societies rather than division.'

Spot on. As Helen Keller, the first deaf and blind person to achieve a university degree, put it, 'The highest result of education is tolerance.'

Yet most education systems prioritise the history of individual nations and states over the history of humanity. As Oxford historian and bestselling author Peter Frankopan told me, 'Governments don't want us to learn that there might be other – maybe even better – ways of doing things.' And within the study of national history, students on traditional courses spend most of their time on conflicts – normally those that their countries won. This is too often at the expense of understanding how political and social systems developed – usually also with blood, toil, sweat and tears – to allow humans to live and work together.

These age-old debates about how we should govern ourselves, how people should treat one another and what we should value have been a huge part of our history, and – for the sake of our survival – must be an even greater part of our future. It starts with a simple statement: history is more than the wars we won.

Some governments have – in fits and spurts – started trying to fill this gap. In 2016, France introduced civic and moral education, aiming to enable pupils to grasp the rule of law, individual and collective freedoms and equality in a democratic society. We must of course be wary of assuming a set of values based solely on Western enlightenment ideas. But in place of education systems that prioritise the teaching of conflict, we do need to understand the political and social systems for coexistence.

Policymakers have wrestled with this challenge, but politicians tend only to grasp it after leaving office, when the temptation to play to populist or nationalist politics has receded. UNESCO's 1996 commission on education, led by Jacques Delors, argued that we should break down that education into four areas: to be, to know, to do and to live together. The final category proved the most controversial at the time, because the report's commission of educators and leaders argued that young people should learn to understand the perspectives of other ethnic, religious and social groups; and to resolve tensions and conflicts.

Somehow it doesn't seem so controversial now, and yet it is proving as hard as ever to introduce.

Some outside governments are also trying to fill the gap. Stories of a Lifetime is a project created to help young people understand each other, 'capturing the moments of yesterday to share, teach and learn from in the future'. Its website exposes children to stories from different countries, including those to which their own countries are opposed. The animation company Big Bad Boo produces entertainment as a tool to teach non-violence, empathy, diversity and citizenship. Its *1001 Nights* series has aired in eighty countries and twenty-five languages. Big Bad Boo has also worked with UNICEF in Jordan to provide refugees with psychosocial support through animated activity books. There are also good online resources already available for people to learn how to de-escalate arguments or solve disputes. These kinds of resources can help, but until the underlying approach to teaching history changes, it will be hard to drive attention and time towards them. So where can we look for examples of how to do that?

Fadi Daou does not look like a revolutionary. He wears a dog collar, and listens quietly and modestly, a slightly quizzical look on his face. You feel that he is consciously choosing to leave space for others to speak. Yet he is on the frontline of the key arguments in the Middle East, and increasingly the world. His Lebanese NGO, Adyan, promotes pluralism, inclusive citizenship and community resilience. It has pioneered a curriculum to teach coexistence among Lebanon's often fiercely divided religious communities. The process of coming up with the curriculum alone took years of painstaking discussion, building trust between sects who had fought each other for much of their existence. What might on the surface appear simple was extraordinarily challenging in the detail. They had to agree not to try to teach recent history, given the extent to which it has been contested by all sides since Lebanon's civil war. But in the end they got to a set of ideas on living together that everyone could agree to teach. This was not just an extraordinary piece of diplomacy. It helps us to see how we could develop an understanding of why and how other communities are different to us.

'We need to shift which "heroes" we learn about,' Fadi told me in a classroom in the Lebanese mountains. 'We started a platform which means hero in Arabic – short stories about people who are going against the current and deserve to be called a hero, in comparison to Islamic State terrorists and others who glamourise violence. We try to give pupils the knowledge and attitudes to coexist in one of the most diverse societies on earth.'

If it can work in Lebanon, it could work anywhere.

The starting point – as my NYU hackathon students constantly emphasised – is to reflect on our shared story. A basic grounding in the history of humans finding ways to live together might include the following highlights.

Language was one of our first great acts of coexistence. Survival of the fittest did not always mean survival of the strongest. Some 50,000–150,000 years ago, beginning with grunts and body language, pioneering humans found ways to communicate and collaborate, allowing prehistoric humans to hunt together and form better-organised social groups. This laid the basis for tribal society. Some ten thousand years ago, life remained transactional, precarious and violent. But the development of agriculture enabled humans to start to build a sense of common interest in the communities in which they lived. The first social networks relied on family bonds and shared enemies.

Raiding and trading became increasingly important features of this new landscape. As empires rose and fell, the routes for passage across continents turned into conduits for the movement of people, goods and ideas about how to live together. Voyages and explorations meant the sharing of more conflict, disease and exploitation, but also the transfer of science, maths, medicine and culture. Each required collaboration between societies, the recording and transfer of knowledge.

For those scholars on the receiving end of the sword, it would not have felt the case, but over centuries the pen really was mightier.

City-states were important early experiments in living together. Societies started to develop and organise in ways that allowed new ideas to flourish. The Greeks invented the early forms of democracy, and Alexander the Great spread democratic ideas from North Africa to India with his armies, a method still much loved by some modern Western democrats. The Romans created and spread systems and a language for organising them across an empire. Violence might have forged empires but it took systems of collaboration to govern and sustain them.

So laws and administration followed these early efforts to organise ourselves in communities. From about five thousand years ago people began to produce written rules and laws in the Middle East, Egypt, India and China. In communities with a surplus of healthy foods, people had more time for discussion and the arts. They began to think about more peaceful ways to settle disputes. The Mongols (late thirteenth century) and Ottomans (fourteenth to seventeenth centuries) became masters at organising communities, developing many of the features of the modern state.

The early empires also saw the benefits of sport to leisure and political stability. The first Olympics brought together men from every city-state in the Greek world. Truces were called for the duration of the games. The combination of the codification of sports and leaps forward in transport and communication would go on to make sport a source of belonging. In the social media biographies that we use to capture our identity, we are now much more likely to express our footballing than our national allegiance. More recently, sport has given many of us a sense of global connection: over a billion people watched the last World Cup final, and many find themselves feeling more patriotic and less nationalist when watching the Olympics.

Nation states were a big experiment in organising ourselves to coexist. From three thousand years ago, communities had begun to establish themselves more firmly on the basis of common aspects of language and culture. In the last millennia, many emerged as the states we know today, sharing systems for living together. The nineteenth century saw the establishment of several of the modern European nation states. Many of the 195 states in existence today were born from conflict, and competition between them has often threatened their coexistence. But they are a relatively new feature of common identity.

The nineteenth and twentieth centuries were a time when rapid technological development led to the industrialisation of conflict and the export (through colonisation) of violence. But this period also saw the next leap forward for living together: the codification of rights and freedoms of speech, the press, religion and globalisation. This is evidently still very much work in progress, but the French Revolution of 1789 helped lead to the US Bill of Rights and eventually (via the horrors of the twentieth century's conflicts) to the 1948 UN Declaration of Human Rights.

This is humanity's greatest text so far, the most powerful, revolutionary and under-appreciated document of all time. The problem is not that we don't understand our duty to our fellow citizens, but that we don't have the will to deliver it. The Declaration established the 'four freedoms' of speech, religion, from fear and from want. They are as powerful and fragile today as when they were first set out. If you haven't read the Declaration of Human Rights, please put this book down and do so.

The middle of the twentieth century also saw the codification of the rules for globalisation, as we developed a sense that we might become collectively stronger by ceding some sovereignty and power to global rather than national institutions. At Bretton Woods, with the end of the Second World War in sight, forty-four nations established for the first time an international banking system, with the World Bank at its heart. The International Monetary Fund (IMF) created a global currency conversion system. These institutions are imperfect, but they are our best effort yet to manage the reality of a global economy that is not going away.

Meanwhile, the effort to develop global values continues. In 2000, 191 nations came together to pledge to end extreme poverty. This pledge turned into the UN's eight Millennium Development Goals, which aim to secure the rights of every person on the planet to health, education, shelter and security. The MDGs – which range from halving rates of extreme poverty to halting the spread of HIV/AIDS and providing universal primary education – form a blueprint agreed to by almost all the world's countries and all its leading development institutions. They and their successors the Sustainable Development Goals are off track, but have galvanised unprecedented collective efforts to make poverty history and – more recently – forge the Paris Climate Agreement.

These were the great landmarks in the history of coexistence. From the vantage point of the 2020s we can see how mass conflict and mass communications have accelerated our understanding of our common identity as a species. But in the meantime we have developed weapons that could wipe us out and technology that could replace us. What can we learn from that history that can help us survive its next chapter?

Most importantly, we can remember that human society advanced one fireside discussion at a time. Far more critical than decisive battles or influential individuals were the millions of unrecorded individual conversations, the millions of unrecorded moments of restraint, the millions of unrecorded simple fragments of wisdom passed on through generations.

Human rights activist Steve Crawshaw has spent a lifetime documenting what he calls small acts of resistance. 'The courage and imagination of these unseen heroes is breathtaking,' he writes. 'And they contribute to extraordinary change.' Visiting me in Beirut, he always encouraged me to look for these in the Middle East and beyond. I still find myself doing it in any news story that seems overwhelmingly grim. Like American television host and educator Fred Rogers, we can look for the helpers. For every 9/11, there are firefighters heading towards danger to protect people. For every pandemic, there are nurses putting in extra shifts and volunteers delivering food to the vulnerable.

And how can we hone the survival skills to become the helpers ourselves?

German physicist Jacob Bronowski's work on the nuclear bomb in the 1940s drove him from the most complex corners of the destructive power of mathematics and science towards a profound understanding of this simple but vital task of how to live together. He concludes *The Ascent of Man*, his powerful televisual account of human development, standing knee deep in a muddy pond at Auschwitz, reflecting on the Holocaust, in which many of his own family were killed. We must never underestimate our potential for evil and violence. Humans are work in progress, or, as Bronowski puts it, 'Knowledge is an unending adventure at the edge of uncertainty.' He talks about the danger of certainty and the importance of recognising that we do not have all the answers.

We need, he exhorts us in the final shot of the series, 'to reach out and touch people'. Our survival depends on making that connection – on whether we are just connected or can truly connect. Understanding how we coexist is not just a point for debate among educators, activists and politicians. It is a survival skill for all of us.

Beyond our responsibility as individuals and citizens, we will also need to apply this practical approach to coexistence to our communities. We can help to do that by focusing on five features of modern society that are frequently sources of division and polarisation: the relationship between host communities and new arrivals; that between old and young; the place of religion; the design of our living environments; and social media itself. They will be some of the great peace processes of the twenty-first century.

Migration is *the* human story. And the migration of the coming decades will be dramatic. We'll face new and unpredicted conflicts in addition to the spillover from long simmering divisions in the Middle East, North Africa, Europe and Asia. Climate crisis and conflict will combine with digitalisation to drive millions to move. The next twenty years will see more migration than in the last fifty, dwarfing what we have seen in the aftermath of the conflicts that exploded from Iraq, Afghanistan and Syria. As Europe has found with the waves of Syrians fleeing Assad's brutality, large numbers of migrants change people's sense of identity and culture.

Often this can be positive, particularly when looking back many years later. But sometimes it can leave native communities feeling powerless and threatened. We will all need the skills to navigate the friction points, to improve communication and to reduce the potential for misunderstanding, discrimination and conflict. Since 1989, as the researcher Max Roser has shown, many more people have died in ethnic or cultural conflict than in wars between countries.

We will need new ways to mediate between communities that are already in place and new arrivals. Much political division is caused by our inability to do this well, with the misconceptions and distrust that generates. One model that might help protect rights and define responsibilities is one in which citizenship and residence are separated. The former gives a right to vote and to access long-term welfare support, for example. The latter gives rights to live and work freely, but without the same obligations and benefits.

With ageing populations, getting better at coexistence also means that we will have to get better at living together across generations. Older people can help by recognising the challenges that young people are facing. The Resolution Foundation found that UK millennials are still only earning the same as those born fifteen years before them were at the same age; that by the age of thirty they are only half as likely to own their home as baby boomers were by that age; that private sector membership of generous 'defined benefit' pensions for those around the age of thirty-five halved for employees born in the early 1980s compared to those born around 1970; and that the UK's ageing population means public spending on health, care and social security is set to rise by £63 billion by 2040. The Resolution Foundation recommends much bolder responses, including social care funding from a tax on property; an NHS levy via National Insurance on the earnings of those above state pension age; practical support and funding for younger workers; and the abolition of inheritance tax. Assuming no government is bold enough to take such measures, we can assume that friction between the generations will increase. This makes these intergenerational arguments a pressing space for fresh ideas, understanding and leadership.

Sadly, much of the history of the major religions seems to have been the story of how they have fought over their differences, instead of the values that they hold in common. The five major world religions (Christianity, Judaism, Islam, Buddhism and Hinduism) all share a sense of community and ritual. Between the three monotheistic religions, most people in the last millennia believed there was one God (God/Yahweh/ Allah) who created the world and spoke to us through prophets. Religions gave us rules and values for living together, such as the Ten Commandments in the Judaeo-Christian tradition. Their holy books – Torah, Bible, Koran – featured many of the same figures and places, each taking up the narrative of the others. Each developed the idea of a life as a choice between good and evil, with an emphasis on loving our neighbours. Each can act as an extraordinary force for coexistence, but can also be abused and manipulated to oppress and divide.

Can we hope that might change as a global community develops? Will religions be part of the solution to polarisation and division, or will they be part of the cause?

In 2011, as anti-government protests rocked Egypt, Christian protesters in Tahrir Square linked hands to create a human wall around a group of Muslims so that they could safely pray. A few days later, during mass in Cairo's central plaza, a group of Muslims joined hands to protect the praying Christians from violence. Earlier that year, Christians had been targeted by Islamic extremist groups in the country, including the bombing of churches in Alexandria on New Year's Day. After the service in Tahrir Square, the crowd of Muslims and Christians chanted 'One hand', celebrating their unity while holding up Korans and crosses.

This need not be restricted to times when religious communities find themselves in violent confrontation. Such lessons could apply to parts of life in which it is politics rather than religion that divides us. Sometimes it might be because of a shared need to find and defend common values, despite our tactical differences. Perhaps we could now imagine the same moment for those people in America who supported and opposed the presidency of Donald Trump. Or those in the UK who supported and opposed Brexit.

The effort to understand our shared need for human dignity and equality of opportunity also applies to the way that we design the places in which we live. Most modern cities have features that unite and features that divide. A public square, library or park can be a leveller, creating a sense of shared space. Yet – whether by accident or design – the towns and cities that developed in the last two centuries are more likely to remind us of the differences between us and those who are living elsewhere in them. The ethnic or national ghettos of the nineteenth century – from Warsaw's Jewish quarter to New York's Irish or Italian quarters – often became a source of friction and a target for violence. The economic or class ghettos of the twentieth century – from London's housing projects to the banlieues of Paris – also acted as a visible reminder of difference and inequality.

We need to design cities of the future that prioritise not just sustainable development but sustainable politics. At their heart should be not only shared spaces but a sense of shared purpose. To be magnetic, a city needs to be better at 'CITY' – culture, innovation, tolerance and youth. The most successful of them find a way to renew themselves by attracting young people. With that combination, cities will create a space for citizens to thrive, flourish, live fulfilled lives, and hold leaders and each other to account. They will convene talent. When we fail to build cities and towns based on tolerance and openness to outsiders, a place becomes repellent, and withers. British philosopher Bertrand Russell was right when he said, 'It's coexistence or no existence.'

Finally, there is social media. We need a forum where humans can debate the collective challenges that we face freely, openly and in less polarising ways. Social media can be this space. But it is currently failing to do that, and in many cases is driving division. With more than a billion users, Facebook and Google are comparable in population size to China and India. Twitter (645 million), Weibo (503 million), Instagram (200 million), Badoo (200 million) and Renren (210 million) are building similar empires, or being merged into the larger ones. These platforms have transformed how we create, live, work and love. They have immense power to hold leaders to account and make governments fairer and more effective; to increase our collective ingenuity; and to show us that we have more in common than that which divides us.

But they also have immense harmful potential to make us more divided, angry, violent or simply apathetic. Twitter lynch mobs can gather quickly and destroy lives and reputations in hours. Social media could become the greatest-ever force for living together in history. Or the cause of the breakdown of much that we have learned about coexistence. That depends to a larger extent than we realise on us, the users.

As the history of coexistence has shown, the way we harness social media, and whether it becomes a runaway train that platforms vitriol and self-interest or a tool for positive interaction, will in the end come down to the millions of decisions we all make in the course of our lives. As Bobby Kennedy famously told South African students in 1966, 'It is from the numberless diverse acts of courage and belief that human history is shaped each time a man stands up for an ideal or acts to improve the lot of others or strikes out against injustice. He sends forth a tiny ripple of hope, and crossing each other from a million different centres of energy and daring, those ripples build a current that can sweep down the mightiest wall of oppression and resistance.' A modern version of this idea might add that each time a human tweets, they can choose to send out a tiny ripple of hope and solidarity or one of fear and division.

If we are to make social media our public square for debate – or if it has already assumed that role in our lives – it is vital to consider how we all conduct ourselves in the forum. My great hope is that the toxicity of social media in the early twenty-first century will come to be seen as like the toxicity of much of the debate after the invention of the printing press. In both cases, it took time to understand this new power. In both cases, the angriest, most extreme voices initially shouted loudest. But the example of the printing press allows us to hope that, in time, we can find ways for the majority to interject, overriding the agitators whose toxic views are not representative, and reclaiming the space for debate, reason, respect and progress.

How do we get there? First, we acknowledge that the only path to a healthier, more democratic mode of social media is by engaging with its current incarnation, but improving behaviour across it. That's a question of considering the tiny ripples of online hope that build currents. It's a question, when looking down the barrel of inequality, poverty, climate crisis and polarisation, of what we share and retweet, of how we platform expertise and fair-mindedness, of how we listen and respond to the voices with which we disagree. We can resist the urge to quote or tweet for attention the appalling view when there is a risk we amplify it by doing so. We can seek out and share the moderate, the consensual, the challenging.

Why not give it a try? There are three things I practise with my students to help us get better at this. Retweet the person you normally disagree with when for once they say something you agree with. Draw attention to someone who brings expertise and nuance to a discussion, especially if they are from a group whose views are underrepresented. Express disagreement with a view without suggesting that the person tweeting or chanting it has no right to do so or is somehow ignorant or malign.

And three things to stop. Resist the urge to tweet the article or the campaign on the basis of the clickbait headline or hashtag. Don't limit yourself to a shallow pool of online influencers. Don't assume that the person whose view you deplored last week will only say deplorable things next week.

Those ripples will start to make it easier to move towards better ways of living together, online and offline. They will help us move faster towards using these new ways of communicating as a means to build not destroy. This effort to coexist requires great courage. As Maya Angelou said, 'Without courage, we cannot practice any other virtue with consistency. We can't be kind, true, merciful, generous, or honest.'

But this bravery is not about being reckless, or about courage in the chest-thumping, silverback sense. It is about a kind of courage we need to value much more highly. This is the courage to be tolerant, to challenge injustice. In Martin Luther King's words, it is 'the power of the mind to overcome fear'. Think of the unknown rebel at the 1989 Tiananmen Square protests, standing in front of the tanks with his plastic bag. Or Rosa Parks refusing to give up her seat in the 'colored' section of the bus. Or Buddhist monk Thich Quang Duc's self-immolation in protest at South Vietnam's oppression.

Who embodies this courage? What does it mean in practice? And how can we be more like them? Bravery will take many different forms. It will not require all of us to stand in front of tanks or set ourselves on fire. But without a surge in the courage to coexist, the twenty-first century will be even more perilous.

'The boat started taking in water and the engine turned off. It was just a little dinghy. I jumped in the water, and my sister joined me; we had to use all the swimming resources that we had. It was a very rough sea that day. But I'm a long-distance swimmer, and I knew I could survive.'

These are the words of Sara Mardini, a Syrian refugee who fled Assad's barrel bombs and the destruction of her home in 2015. She and her sister, who in 2016 would go on to swim at the Rio Olympics, spent almost four hours dragging and pushing the boat in which other refugees were making the dangerous crossing to the island of Lesbos in Greece. She describes them as the brave ones. 'They couldn't swim like us, but they worked together.' She and her sister now volunteer as rescue workers among the refugees still arriving. In 2018 Sara was imprisoned for over a hundred days for this work. 'Normally we reward lifeguards, firemen and doctors who save others,' she says. 'I think we need to go back to the basic things kids learn at school – to love others, to take care of your neighbours.'

Sometimes this kind of courage simply requires us to be more humble. 'I have seen how you can't learn anything when you're trying to look like the smartest person in the room,' wrote the novelist and poet Barbara Kingsolver. Winston Churchill agreed: 'Courage is what it takes to stand up and speak; courage is also what it takes to sit down and listen.'

As with all the survival skills, we can train ourselves. That applies just as much to developing the courage to coexist as to learning a language. Here are some more exercises that might help.

1) OBSERVE

Who are your role models? As I've tried to do in this chapter, write down your own profiles in courage and coexistence. I believe 'X' is brave. Here's why. And here's what I learn from that.

2) PUT ON YOUR OPPONENT'S SHIRT

The man who managed to do it best in the last century was Nelson Mandela. His extraordinary effort clearly demanded much patience, resilience and creativity. But there was one more vital feature, which can be learned.

The 1995 Rugby World Cup was the first major sporting event to take place in South Africa following the end of apartheid and Mandela's election as the country's president. With the South African team facing the formidable New Zealand All Blacks in the final, Mandela chose to wear the Springbok jersey, a symbol of repression for many black South Africans. This signalled to blacks that it was fine to get behind the national team, and it signalled to whites that he was their president too, whatever the past. 'When Nelson Mandela walked into the changing room wearing that jersey,' said winger Chester Williams, 'it was done. We had to win that game. It changed the attitude and spirit of the team – and the whole mindset of the nation.'

South Africa did win the rugby that day. As he collected the trophy from his delighted (and no doubt relieved) president, team captain Francois Pienaar said, 'We did not have 63,000 fans behind us today; we had 43 million South Africans.'

Much could be achieved in modern political and social media debate if people were more often willing to occasionally put on their opponent's shirt. This might mean taking a knee or simply listening. But also to be ready to accept that there are moments when others might choose not to sing our song, salute our flag or tolerate our choice of statue. In those moments, the reflex is often to assume that there is only one way to think, that an issue is a case of right and wrong, that our opponents are somehow evil. Yet in doing so, we become part of the problem.

You can test and practise this by taking a moment in history or a story from a news website. Decide who you instinctively sympathise with, then try to see it from the other side's perspective. Try writing their opening statement in a debate.

3) MORE IN COMMON THAN DIVIDES US …

List five things you have in common with your adversary, opponent or the person who just pushed in front of you. Imagine their background, day or mood. Take yourself out of the equation. What action can you take to show an understanding of their outlook? And when you need to take a stand, what is a quiet yet powerful way to do that? You might be the Brexiter who calls out the trolling of a Remain campaigner. You might be the Democrat who points out that it doesn't help to label all Republicans as deplorables. We can practise this practical empathy until we no longer notice we are doing it.

4) TAKE TWENTY SECONDS OF INSANE COURAGE

There is so often something we know we need to say or do, and we hold ourselves back. Experiment with taking those twenty seconds. It need not necessarily be a skydive. It can be telling someone you love them. Making a speech or an intervention. Standing up for yourself. What's the most courageous thing you can do today? What's stopping you?

5) SHOW VULNERABILITY

Seek out moments at work, home or in the community to simply ask questions and admit weakness. It is much easier than it sounds, and much more powerful than you can imagine.

6) WRITE YOUR OWN MALALA SPEECH

What would you be prepared to die for? How are you demonstrating the change you want to see?

7) TEACH COURAGE

How are you helping others to be braver? 'Keep your fears to yourself but share your courage with others,' counselled Robert Louis Stevenson. You can help kids to stand up for themselves by knowing when to stand back; listening; coaching; practising with them; setting an example; and having a democratic household.

One final Northern Ireland story. During one intense all-night negotiation, we were challenged by the fact that one of the Unionist leaders refused to face his Sinn Fein counterpart, because of a horrific act of violence against a family member. The Unionist entered the room, but he sat with his back to the other politician. At that moment, I had to share with the room – because for security reasons no one was allowed to take in a phone – the shocking news that the Sinn Fein politician's mother had died. We of course offered an immediate adjournment. He sat quietly for several moments, a tough man wrestling with his emotions. 'No. My place is here,' he said. The atmosphere in the room changed, and while the two sides still sat in fierce and fundamental opposition, the shared experience of loss, struggle and frustration became blazingly apparent. The Unionist turned his chair round and the negotiation proceeded, face to face.

Our survival sometimes requires us to turn that chair round, to recognise that moment of common humanity. It requires us to be kind.

7

How to Be Kind

Kind words can be short and easy to speak,
but their echoes are truly endless.

Mother Teresa

Kenyan athlete Abel Mutai was just a few metres from the finish line. But he became confused by the signage and stopped, thinking he had completed the race. A Spanish athlete, Ivan Fernandez, was right behind him and realised what was happening. Fernandez started shouting at Mutai for him to continue running; but Mutai didn't understand Spanish so didn't know what Fernandez was saying. So the Spaniard stopped running and pushed Mutai to victory.

Asked afterwards by a journalist why he let the Kenyan win, Fernandez replied, 'I didn't let him win, he was going to win.' The journalist insisted, 'But you could have won!' Fernandez looked at him as though he was crazy. 'But what would be the merit of my victory? What would be the honour of that medal? What would my mum think of that? Values are transmitted from generation to generation. What values are we teaching our children? Let us not teach our kids the wrong ways to win.'

The reality is that we see great examples of kindness like this all around us. We just need to open our eyes to them.

This is not a chapter about being kind for the sake of being kind, or nice for the sake of being nice. As Canadian psychiatrist Marcia Sirota has put it, 'At the root of extreme niceness are feelings of inadequacy and the need to get approval and validation from others.' This is instead a chapter about recognising kindness as a genuine, hard-won, hard-done, survival skill. In 2019 I was privileged to help host naturalist Jane Goodall when she visited the Middle East. I watched schoolchildren in Abu Dhabi hang on every word she said, her voice seasoned but forceful, every sentence carefully chosen. She spoke about kindness as being part of our DNA. 'We have the choice to use the gift of our life to make the world a better place – or not to bother. We have so far to go to realise our human potential for compassion, altruism and love.'

So why does kindness figure in a list of survival skills?

First, because without it we will not be able to reduce the inequality I outlined in the introduction, described by the World Economic Forum as the greatest risk humanity faces. Get this wrong and we face a dystopian era in which those of us fortunate enough to have more resources will find ourselves in smaller and smaller gated communities, trying to hold back an angry, hungry world. Wealthier people might console themselves that the consequences of pandemics, crime, extremism, unemployment and injustice hit poorer communities first. These things normally do. But history suggests that no one stays immune for long.

We need to find kind people, or greater kindness in ourselves. David Orr, professor of environmental studies and politics at Oberlin College, has argued that the planet does not need more successful people in the traditional sense: 'But it does desperately need more peacemakers, healers, restorers, storytellers and lovers of every kind. It needs people who live well in their places. It needs people of moral courage willing to join the fight to make the world habitable and humane. And these qualities have little to do with success as we have defined it.' We need therefore to redefine, or perhaps rediscover, what it means to live a successful life.

Second, and more selfishly, kindness is a survival skill because it makes us happier. The science backs up the Dalai Lama: 'If you want others to be happy, practice compassion. If you want to be happy, practice compassion.'

The UN-sponsored World Happiness Report looks at the factors that can contribute to happiness: income, health, life expectancy, generosity, social support, freedom and lack of corruption. It concludes that inequality of wellbeing is more important than income inequality in explaining average levels of happiness. We are happier to live in societies with less disparity in their quality of life, where we have greater trust in each other and our institutions, and we care about the welfare of others. In these communities, we are readier to extend that kindness to other countries and future generations. Social environment is a huge contributor to our wellbeing, especially having someone to count on, a sense of freedom to make key life decisions, and feeling part of a community. People in communities that have higher rates of trust between members are much more resilient in the face of a whole range of challenges to their wellbeing, such as illness, discrimination, fear of danger, unemployment and low income. This explains the higher recorded happiness levels of the Nordic countries who score highly on trust such as league leaders Finland and runner-up Denmark.

An eighty-year Harvard University study of human well-being has confirmed that the single most important factor for happiness is indeed relationships. Surveys by the University of Chicago's National Opinion Research Center found that those with five or more close friends were 50 per cent more likely to describe themselves as 'very happy' than those with smaller social circles. A walk raises our mood by 2 per cent, but it raises it by almost 10 per cent if we take that walk with someone else. Company also makes activities like commuting and queuing less miserable. And people in happy, stable, committed relationships tend to be far happier than those who aren't. Kindness is the most important predictor of satisfaction and stability in a marriage. No one pretends that any of that is easy, but that's another book. Or maybe several shelves of them.

Education pioneer Sir Anthony Seldon has shown that when schools put kindness at the heart of curricula, students learn to be open-hearted and open-minded. Launched in 2014, his International Positive Education Network aims to foster emotional intelligence and wellbeing in young people. Each year it convenes tens of thousands of educators who are experimenting with ways to develop these attributes. Most of the progress is being driven by teachers and learners rather than governments. Yet kindness and empathy are not something to develop alongside education: they are central to our education. In twentieth-century models, logic was rated highly, so our children were taught to memorise times tables and do handwriting. We should now be developing kindness and empathy with the same rigour and focus.

Third, we need to be kinder because it makes us healthier. Kindness reduces anxiety and stress. Meditating on a compassionate approach to others shifts our thinking to the left hemisphere of the brain, a region associated with happiness, and it boosts immune functions. It even changes the chemical balance of our heart: the hormone oxytocin causes the release of nitric oxide in blood vessels, which dilates those vessels and reduces our blood pressure.

Fourth, we should learn kindness because Jane Goodall was right – it is in our DNA. Darwin's survival of the fittest is often wrongly interpreted as being about selfishness and competition. But Dacher Keltner, author of *Born to Be Good: The Science of a Meaningful Life*, has argued that the innate power of human emotion to connect people with each other is the path to living the good life. He interprets evolution as having 'crafted a species with remarkable tendencies toward kindness, play, generosity, reverence and self-sacrifice', which in turn enhance our survival. 'These tendencies are felt in emotions such as compassion, gratitude, awe, embarrassment and mirth. Our capacity for caring, for play, for reverence and modesty are built into our brains, bodies, genes and social practices.' As Darwin argued, we are a profoundly social and caring species.

Finally, kindness is a survival skill because it makes us more successful professionally. In a study of its most successful teams, Google found the key common factor to be kindness: groups that took care of each other were the most innovative, productive and happy. And this was not just a hunch. According to the journalist Charles Duhigg, the company 'scrutinized everything from how frequently particular people eat together (the most productive employees tend to build larger networks by rotating dining companions) to which traits the best managers share (good communication and avoiding micromanaging)'. The initiative, called Project Aristotle, gathered Google's top statisticians, psychologists, sociologists and engineers. The conclusion? 'No one wants to put on a "work face" when they get to the office. No one wants to leave part of their personality and inner life at home. But to be fully present at work, to feel "psychologically safe", we must know that we can be free enough, sometimes, to share the things that scare us without fear of recriminations. We must be able to talk about what is messy or sad, to have hard conversations with colleagues who are driving us crazy.' They concluded that this meant shifting the focus away from efficiency alone. 'We want to know that work is more than just labour.'

If we invented a GPS to guide us through life, it would be consistently directing us towards being kinder to ourselves and others. Maybe it already is. Activation of the vagus nerve is associated with feelings of kindness and a sense of common humanity. It makes us more likely to be altruistic, grateful and loving. Psychologist Nancy Eisenberg has shown that children with higher vagus nerve activity are more likely to give, to cooperate. Altruism is an evolved survival instinct.

That evolution in our kindness was hard won. Many of us think that we have a set of fairly straightforward values, the product of nature and nurture, accumulated from family and experience. But in reality our inheritance is far more complicated, and shaped by historical context. Our earliest stories and epic poems – Mesopotamia had the *Epic of Gilgamesh*, the Greeks had Homer's *Iliad*, the Icelanders had the *Eddas* – were more likely to have been motivated by fate and vendetta than kindness and compassion. Of the early philosophers, for every Socrates who argued for virtue there was a Plato or Epicurus who countered that it was more important to ensure the smooth running of the city or to focus on pleasure. The ancient religions developed the golden rule: do not do to others what you do not want done to yourself. This became the centrepiece of Jewish and Christian ethics. In the Middle

Ages, Thomas Aquinas combined biblical ethics with the ideas of Aristotle to argue that kindness was worthwhile because it made society more resilient, not just for its own sake or because it increased the odds of paradise rather than eternal damnation.

As articulated in the eighteenth and nineteenth centuries by Immanuel Kant, Jeremy Bentham and John Stuart Mill, humans increasingly believed in the idea of the common good. The horrors of the Second World War drove us to codify a balance between our rights as individuals and our responsibilities to others. Much modern debate – such as on the limits of free speech on social media platforms – takes place in the contested areas between the two.

So our understanding of kindness has changed significantly over time, and continues to change. But we have inherited a powerful sense, developed over millennia, that kindness is good both for us and for wider society. However, unlike maths, science or a foreign language, we haven't created time and space to learn and practise how to be kind. As a result, we risk delegating our understanding of this vital skill to Walt Disney. And we risk relegating this potential superpower to becoming something that we used to hear about in places of worship.

Central to an ability to be kind is the increasingly important survival skill of empathy. Can we learn both empathy and the means to manage and use it more effectively? Doesn't it simply come with time and experience? During the pandemic, a leader like Jacinda Ardern of New Zealand demonstrated through her policies and the way that she communicated them how empathy can enable better decision-making and acceptance of those decisions. One of Joe Biden's most attractive qualities as US president has been to share people's pain.

But could you teach more authoritarian or narcissist leaders empathy? When I put this to neuroscientist Tara Swart, she cited examples of hardened criminals who had learned remorse and empathy. 'Many of the very successful people I have worked with had a sort of minor childhood trauma,' she says. 'Particularly if this happens from around the age of five to seven years old. A mild to moderate form of stress or trauma can lead to a virtue. If you understand how your emotions work then you've got a massive head start.'

But did this mean another area for parents to screw up? Alarmingly, empathy and the ability to see your own filter seems to be declining in younger people, and individualism is rising. Thankfully, Tara was more encouraging.

'Children need to feel safe and loved and if you do that then they're in the environment to learn all of those other things,' she says. 'Partly through your role modelling but also at school, but mostly through the ability to really get to know themselves. We don't need to be forcing our kids to play musical instruments. The most important things we need to be teaching children and growing pathways for in their brains are being able to understand and manage their own emotions and relate appropriately to other people's emotions.'

This doesn't have to be part of formal education alone. Most of us have learned empathy from life, family and popular culture. While I have argued that we can't subcontract the teaching of kindness to cinema, films can help. Take *Inside Out*, a Pixar movie that tried to help us understand our own emotions. The story takes place inside the head of an eleven-year-old girl, as her emotions – joy, sadness, anger, fear and disgust – jostle to help her make sense of the world. The film was developed with scientific insights on how our minds work. As a result, it has helped young and old people develop a sense of happiness being more than just joy, and not something that can be constant. Or that sadness can contribute to our well-being: the Sadness character is even the heroine of the film.

Practising kindness also means being kinder to ourselves. During the lockdowns of 2020 and 2021, I regularly asked my students how they were coping. Over half said that they had experienced mental health challenges. The enforced distancing combined with wider anxieties over their health, their families or their studies. A toxic mix for many. Even without the pandemic, these students are not the only ones. One in four Brits now report experiencing some kind of mental health problem. There is a growing sense that something is missing. We have this nagging feeling that life should be a bit better, that we should be a bit happier.

The science of happiness is worth billions. And it is highly contested terrain. Yet I've found speakers at happiness conferences to be among the most miserable around. Like financial advice, it is a massive industry sustained by people's need for guidance. And, as with financial advice, strip away the soundbites and marketing and the guidance we need often turns out to be based on something very simple: common sense.

Wellbeing is a journey not a destination. We can and should work at it, and there are tools and skills that can help us. We can also be kinder to ourselves by recognising that many of the factors influencing our happiness are beyond our control or agency. Some of these require the more systemic responses – to confront injustice, oppression and inequality – which we looked at in the last two chapters.

There is an important reason why growing numbers of us feel that we lack the practical skills to be kind to ourselves: we were never taught them. When my NYU research team asked students from thirty countries what they were missing out on in their education, over two-thirds stressed health education, from emergency first aid to diet and exercise. Every single group mentioned mental health. In an increasingly stressful and 'always-on' world, they felt they lacked the skills to cope. They also worried that they would leave formal education with no idea of how to manage their finances, increasing their stress levels and reducing their ability to live a life of purpose, flexibility or control.

So how are kindness and empathy really learned? As with the other survival skills, it is by observing, learning, practising and teaching. Like working out in a gym or studying for an exam, this needs real time and consistent dedication. Here are seven ways to try.

1) LOOK FOR KINDNESS

Observing kindness could start with looking for examples of it from history. They are everywhere. Take Sousa Mendes, Portugal's consul in Bordeaux during the Second World War, who saved the lives of thousands of Jewish refugees and others fleeing the Nazis by issuing travel visas from occupied France. According to his son, 'He strode out of his bedroom, flung open the door to the chancellery, and announced in a loud voice: "From now on I'm giving everyone visas. There will be no more nationalities, races or religions."' Thousands escaped and most sought to cross the Atlantic. Among them were artist Salvador Dalí, filmmaker King Vidor, members of the Rothschild banking family and the majority of Belgium's future government-in-exile. Mendes was expelled from Portugal's diplomatic corps and left without a pension.

Harriet Tubman rescued at least three hundred people from slavery in the 1850s. Born a slave and having worked from the age of five, she fled alone in her mid twenties to Pennsylvania, the neighbouring free state, helping to test a new 'underground railroad' that was being used by slaves and abolitionist sympathisers to escape. Despite the threat of harsh penalties, she made as many as nineteen trips to rescue other slaves and deliver detailed instructions on how to escape.

In the slums of Kibera on the outskirts of Nairobi, Ernest, a teacher, tells his pupils the story of the old man who sees a child looking for starfish on the beach and carrying those he finds back to the sea. The boy explains that the washed-up starfish will die before the tide returns. 'But there are tens of thousands of the creatures,' worries the old man. 'You won't make much of a difference.' The boy picks up another starfish and returns it to the water. 'It made a difference to that one.'

2) LOOK AFTER THE PHYSICS

Covid-19 and the lockdowns that followed have been a reminder of how important it is that we are kind to ourselves: that we stay in good mental and physical health. Our bodies and minds are living, breathing organisms, after all. They rely on exercise, sleep, oxygen, a decent diet.

We already feel bombarded and overwhelmed by advice on what to eat. Ed Walsh, a science and education adviser, has developed material to help young people – and all of us – discern which advice is most credible. You can do exercises that help you distinguish between faddish newspaper head-lines – 'Olive Oil Stops You Wrinkling!'; 'A Glass of Red Wine a Day Extends/Reduces Your Life Expectancy!' – and the actual research findings. This matters because we need to see through the fads and clickbait and focus on what actually influences our survival chances. Obesity kills three million humans a year. The *British Medical Journal* has predicted that the biggest killers in 2030 will be heart disease and strokes, yet we eat twice as much salt as we need. Almost half of current deaths are caused by bad lifestyles.

Switching to a more plant-based diet doesn't need to mean a life of abstinence and misery. It doesn't even mean you have to give up meat entirely. Just that you ensure that most of your calories come from vegetables, grains, fruits, beans, seeds and nuts. It helps your body fight infections, reduces the risk of cancer and the inflammation that can lead to diseases like arthritis, helps you get to a healthy weight and gives you the fibre to lower cholesterol and keep blood sugar stable.

We can also slow down our eating. Former Arsenal player Ray Parlour has said the team's success in the late nineties was partly due to manager Arsène Wenger's insistence on 'chew to win'. Increasingly, we find ourselves doing something else while eating – checking our timelines, watching Netflix, driving, working. This mindless eating may be making us more obese and less healthy, according to Dr Lilian Cheung, a Harvard nutritionist. Three ways we can be more attentive to our food are to plan, pause and ponder:

- Take time to prepare menus and consider the nutritional value of our shopping lists, rather than impulse buying in the wrong aisles.
- Come to the table with an appetite and pause to appreciate the food and the work that got it there.
- Start with a small portion and eat slowly; be more attentive to the taste, smell and look of the food, picking out ingredients; take small bites and chew food 20–40 times.

Apart from the general smugness that you may feel and the pleasure of a less expensive shopping cart, microbiologist Professor John Cryan believes that new research combining his field with neuroscience, genetics and the brain-imaging of animals and humans is showing the link between our gut health and our mental health. At a conference in Dubai on happiness he told me that this could be the start of 'the psychobiotic revolution. The microbiome makes a major contribution to brain development in the early years of our lives. We are 99 per cent microbes. In the twentieth century, the major focus of microbiological research was finding ways to kill microbes with antibiotics. This century we will appreciate that a healthy gut may be essential to happiness as well.'

Taking care of the physics inevitably takes us to exercise. Getting more regular exercise must be the most overused and under-delivered New Year's resolution. The market has seen that we need help: in 2021 the fitness industry was estimated to be worth over $100 billion. Wearable technology that can track our every move and heart beat means that we now have almost an unlimited source of personal fitness data. My watch can sense when I am having a heart attack and call an ambulance. My last words might be 'Hey Siri'.

I don't want to add to the bombardment of messaging about the need to exercise to lose weight, to look better, to get more Instagrammable abs. According to the NHS, some form of exercise each day – two to three hours a week – means up to a 30 per cent lower risk of depression. A good yardstick: are you watching more sport than you're doing?

The physics of being kinder to ourselves is not just about food and exercise. George W. Bush was often criticised for being lazy. But when I asked him what the key was to becoming US president, he gave a surprising answer: sleep. Even at the most stressful moments, he insisted on a minimum of seven hours to recuperate. Perhaps Bush is not the most consistent example of smart decision-making, but this is a useful counter to the approach many take to their work and that I experienced in Number 10 and much of my diplomatic life – that exhaustion is the way to show you are trying. There is a book to be written on the great mistakes of history made because of a lack of sleep. I hope that someone writes it, after a good night's rest.

So we need better sleep. Over two-thirds of students are getting insufficient rest, with fewer than 10 per cent sleeping the nine hours recommended by medics. The stats are worst among female and black students, and those in their late teens. When we lose sleep, it's harder to focus and pay attention. This affects our performance and productivity. It slows our reaction time: a National Sleep Foundation poll found that nearly a third of drivers said they have nodded off or fallen asleep while driving. Sleep feeds creativity and new ideas. Memories are reactivated, connections between brain cells are strengthened, and information is transferred from short- to long-term. Sleeping after we learn new information helps us retain and recall that information later.

Sleep is particularly important during adolescence, a phase of rapid growth and development. Most young people are therefore operating below par, which accounts for much daytime sleepiness, depression, headaches and poor school performance.

How can we get better sleep? Again, it is not rocket science. The secret, according to the NHS, is to keep regular sleep hours; create a restful environment (dark, quiet and cool); move more during the day. Experts say we shouldn't force ourselves to sleep if we have been lying in bed for too long. Get up, do something soothing and try again once you feel more sleepy. Write down your worries. Do anything you can to help put your mind at rest: write a diary, or make a to-do list for the next day. Try and empty your head before going to bed. And of course, reduce the caffeine and alcohol, especially near bedtime.

The actor Goldie Hawn has long practised mindfulness. But in Los Angeles she told me that after 9/11 she realised that this was a campaign that needed to be felt far beyond California, where she lives. Through her foundation, she is now translating a mindfulness curriculum into more languages. Pioneering educators in the Middle East and beyond have recognised its potential. Gulf schools are now teaching mindfulness alongside maths and science. She told me how we can develop practical tools and exercises to help young people manage their emotional wellbeing, and to be kinder. 'Parents associate meditation with some form of religiousness, whereas a brain break does not have this connotation,' she said. She returned to the attacks on America as the watershed moment for her in realising the risks if we get this wrong: '9/11 made me realise why children are referred to as "soft bullets": the way in which we approach their education can leave space for radicalisation.' Hawn herself started meditating in 1972. She said that it helped her understand and differentiate herself from those that adored her. 'They didn't know me. It was a long journey, not a one-time stop.'

Being kind to ourselves in these ways also contributes to the health of our brains, an important issue in an ageing population. As a neuroscientist and bestselling author, Tara Swart understands the brain. But unlike most neuroscientists, she can explain it. She shared with me her sense of urgency on the need for us all to take better care of our brain health, particularly amid the sensory bombardment of the digital age. 'Our children are growing up in a world where they are constantly multitasking,' she says. 'We've already seen that our memory and concentration centres in the brain have started to shrink. Most people our age would know the phone number of the house they grew up in and probably the numbers of the friends they went to school with. But many people don't know their own spouse's or child's phone number because you have it at the click of a button. So we don't remember things because we don't need to in this world and that really leads me to the crux of the problem around education: it hasn't changed at all to keep up with the way the world and our brains are changing.'

When we recall memories, or join the dots between ideas or experiences, we create a connection in the brain: this is called neuroplasticity. Essentially, our brains are changing all the time, in response to people we meet or emotions we feel. The process is fastest up to the age of twenty-five. In our first eighteen months, we go through huge changes, and then during our teens we prune away the nerve cells and connections we don't need. This is why teenagers need so much sleep. We shouldn't be waking them and packing them off to school: we're disrupting their learning.

The latest research in neuroscience reveals that there is a hierarchy in the things that are most important to our brain health. At the top are a sense of belonging, a strong sense of personal and work identity, and positive meaningful social relationships. Then come aerobic exercise, quality sleep, diet, hydration and time away from the devices. These allow the brain to rest and recoup its resilience and use its resources for optimisation rather than for coping with the toxins of stress.

Given the science on sleep, exercise and diet, we can change our routines to get the extra hour we need in bed or in our trainers, or the extra plants we need on the plate. We can get closer to nature and be at greater peace with it to be at peace with ourselves. All of this will help our brain health too, boost our survivability, and might even make us better to be around.

3) COUNT YOUR BLESSINGS

We can start being kinder to ourselves by counting our blessings. Psychologists Robert Emmons of the University of California and Michael McCullough of the University of Miami found that those who kept journals recording the things they were grateful for ended up feeling happier, healthier, more energetic and more optimistic than those who didn't. To assess global happiness levels, the polling organisation Gallup uses the Cantril ladder. This asks respondents to think of a ladder, with the best possible life for them being a 10, and the worst possible life being a 0. They are then asked to rate their own current lives on that 0 to 10 scale. You can do the same to keep track of your responses over time, and ask yourself what is causing the change.

You can establish rituals that help. Every morning the Dalai Lama reflects that 'Today I am fortunate to have woken up, I am alive, I have a precious human life, I am not going to waste it. I am going to use all my energies to develop myself, to expand my heart out to others, to achieve enlightenment for the benefit of all beings, I am going to have kind thoughts towards others, I am not going to get angry or think badly about others, I am going to benefit others as much as I can.'

You don't have to be the Dalai Lama to do this. What's your equivalent?

Given the centrality of relationships to our happiness, we can also list the relationships that matter most to us. And make a point of telling the people on that list why they are there.

4) PRACTISE KINDNESS

One way to do this is to keep a record. When you believe in someone, tell them directly. Aim for five positive interactions a day. Log them to hold yourself to account (until it becomes a reflex). Take a few minutes before you go to bed to reflect upon your day. Think about the people you met and talked to, and how you treated each other. How well did you do? What could you have done better? What did you learn? The intentionality helps.

This can help us develop a more methodical approach to kindness. In *Kindness: The Little Thing that Matters Most*, Jaime Thurston has identified fifty-two simple things you can do to spread kindness, from being a seat vigilante to being kind to unkind people. Bernadette Russell pledged to be kind to a stranger every day for a year. In *The Little Book of Kindness*, she documents what she learned and offers suggestions to help readers practise being kind.

We can also flex the kindness muscle by genuinely celebrating someone else's success. This is harder than it sounds. It is perhaps not too difficult to recall a moment when you might have felt a tinge of envy, or downplayed the reasons for the success. If genuinely celebrating it feels like too big a jump, at least think before you speak. As my grandmother used to say to us, 'If you have nothing nice to say, say nothing.'

5) PAY IT FORWARD

Another way to learn to make kindness a default approach is to practise paying it forward. 'Act with kindness, but do not expect gratitude,' advises Confucius. Do something kind that no one sees. The Pay it Forward Movement encourages everyone to do three unprompted good deeds for three different people – such as giving someone an umbrella when it's raining or paying for a coffee anonymously – and asking nothing in return except that the person 'pay it forward'. The movement has spread around the globe and even has a book and a film named after it. Pay It Forward Day is a worldwide celebration of kindness every year on 28 April. But you're allowed to try it every day.

6) TEACH KINDNESS

We can also learn kindness by cultivating empathy in young people. Set an example. Harvard's Graduate School of Education has come up with five tips for how to do this: empathise and model empathy for others; make caring for others a priority and set high ethical expectations; provide opportunities for children to practise empathy; expand your child's circle of concern; help children develop self-control and manage feelings effectively. As with all the survival skills in this book, the act of teaching helps to cement the skill. Truly empathic people help those around them to become more empathic.

7) KNOW YOUR FILTER

Becoming more kind also requires us to become more self-aware, to understand our filters: the lens through which we see the world. The starting point is to genuinely understand an obvious but often overlooked truth: ours is not the only way. We are just one perspective.

Take this poem by Brian Bilston:

Refugees

They have no need of our help
So do not tell me
These haggard faces could belong to you or me
Should life have dealt a different hand
We need to see them for who they really are
Chancers and scroungers
Layabouts and loungers
With bombs up their sleeves
Cut-throats and thieves
They are not
Welcome here
We should make them
Go back to where they came from
They cannot
Share our food
Share our homes
Share our countries
Instead let us
Build a wall to keep them out
It is not okay to say
These are people just like us
A place should only belong to those who are born there
Do not be so stupid to think that
The world can be looked at another way

(now read from bottom to top)

Perspective matters. I introduced you in Chapter 1 to Noah Raford, Dubai's futurologist, who was helping me to understand how we can get better at anticipating – though not predicting – the future. The most fascinating part of our discussions was actually where our two worlds of diplomacy and technology met. I had talked about an ability many diplomats have to zoom out from a situation, and to understand the contexts and motivations of those around them. We will look at this skill in more detail in the chapter on being global. It is a form of practical empathy.

Noah was adamant that it is also the key skill of great futurologists. 'But the real ninja skill is to go a step further,' he said. 'Not just to recognise that everyone has a filter or lens, but to genuinely acknowledge and understand your own filter.' Just developing an awareness that you have that filter is a start. What are the assumptions we build into how we view the world? Where do they come from? Are they helpful?

Noah thinks you can then hone that awareness of how our filters work. 'Develop hypotheses about how the world *might* work and then explicitly test them. Seek out differences. Observe it. The things you can't explain, that prove you wrong, that make no sense to you. Recognise that for the person who holds that view, it makes complete sense to them. Those become the sources of new collaboration and learning. Once you really understand your own filter you can adjust your way of being in the world with much greater confidence. Tech can augment this skill. But it is 90 per cent human.'

Like the other survival skills, this extreme empathy and self-awareness can be observed, studied, learned, practised and taught. But beware. Empathy is exhausting. Empaths are highly sensitive: 'If you want heart, empaths have got it,' says Judith Orloff, author of *The Empath's Survival Guide*. Sometimes empaths have too much heart. They absorb other people's emotions and moods, good and bad. They rely heavily on intuition and their gut feelings about people. Their sensitivity makes them easily drained by those who are angry or anxious, and by noise.

This extreme empathy – attempting to navigate situations while consciously observing your filter and the filters of others – is even more demanding. But if we all get better at it, we will boost our ability not just to live together but to reason together.

Four exercises I do with my students at NYU might help.

1. What's my filter? Write down or discuss your reaction to a media story or film. What made you particularly excited, sad or angry? Exchange your conclusions with a partner or in a group. Try to understand why others had different reactions. Rather than attempting to convince them that your response was the correct one, try to listen to their explanations and imagine a different way of seeing the world. We all have biases, the result of our nature and nurture. Many are unconscious. What's their filter? And what's yours? As the policymaker and thinker Cass Sunstein has shown, people's views become more extreme the more like-minded are those they spend time with. If we spend more time with people with divergent views, we develop empathy and seek collaboration.

2. Discuss a negative emotion. This is hard. It takes time and energy. Identify and practise language that can express your reaction in a way that is not adversarial. Not 'You made me angry' but 'When you said that, I felt angry'. Not 'You upset me' but 'I felt upset when that happened'. Try to separate their intent from their impact.

3. Seek honest feedback. This is also uncomfortable. Most people don't give balanced, honest criticism and so we can often blunder through life or work unaware of simple adjustments we could be making. I

find that one way of addressing this is to invite anonymous feedback, for example via a colleague or through written comments. Make clear that you want areas for improvement that you might not spot, that they are helping you by being honest. Sometimes it makes it easier for others to give that kind of feedback when you invite them to give three positives and one challenge. Of course, it is human nature to then focus on the challenge.

4. To try to understand patients, especially those whose views and actions they find troubling, psychologists use a method based on five 'P's. First, they consider factors that might be 'pre-disposing'. What is the family or social background of the person? Was there an early experience in their lives that shaped them? Then they consider 'presenting' factors. Is there something in the person's more immediate circumstances that has made them act in this way? Perhaps a change of job or relationship, or the loss of a friend. Third, psychologists consider 'precipitating' factors. This could include online activity, drugs, trigger events. For difficult cases, they look at 'preservation' factors. What makes the anger continue? What are the habits or emotions that sustain troubling behaviours? Finally, 'protective' factors. What makes it likely that the person can change? Find that and you can start to make progress with them. That sort of framework can make it easier for us to understand the filters of those with whom we disagree. Try it on a stubborn grievance or corrosive relationship. Even more challenging, applied to ourselves, the same questions might also help us to become more self-aware.

When we experience heroic kindness and empathy, the Darwinian instinct for human solidarity is energised. We feel it. One of the most extraordinary pieces of television that has ever been made is Esther Rantzen's 1988 *That's Life* interview with Nicholas Winton, the British banker who organised the escape from the Nazis to Britain of 669 children, mostly Jewish. In the programme, Rantzen describes how he arranged safe passage and found homes for the refugees in the UK. The operation later became known as the 'Kindertransport'.

'Does anybody here owe their life to Nicholas Winton?' she asks the studio audience. After a brief pause, the woman next to him slowly lifts her hand. There is an incredible moment as he realises that he is sitting next to one of those he saved. He turns to her with emotion. Time seems to slow as you feel the human connection across the decades, the human impact of his courage.

And then, as Nicholas dabs his eyes with a handkerchief, stands up and looks around him, the whole room rises. It is hard to watch the video without feeling the wave of deep solidarity. Never let anyone tell you that kindness is not a superpower.

With this more conscious effort to understand our filters, to learn and teach kindness and empathy, we can develop the next advanced survival skill: to befriend the machines.

8

How to Live with Technology

So it is the robots you say you fear. Why fear
something? Why not create with it? Why not build
robot teachers to help out in schools where teaching
certain subjects is a bore for EVERYONE? ... I am not
afraid of robots. I am afraid of people, people, people.
I want them to remain human. I can help keep them
human with the wise and lovely use of books, films,
robots, and my own mind, hands, and heart.

Ray Bradbury

With radical kindness we will find new ways of working and
living together in the face of the collective challenges that lie
ahead. But what does this mean for our relationship with
technology? Should we fear it, and defend ourselves? Or will
we find ways to coexist with the machines, maybe even to be
kind to them too?

The pessimists are right to warn us to be vigilant about the
next phase of technological change, as they were right to

warn of the negative and unintended side effects of previous technological breakthroughs. Why? The pace of this wave of overlapping changes could mean we become increasingly addicted to and enslaved by technology. We might find the cult of the amateur prevails to a greater extent, with culture dumbed down and imagination, idealism and creativity increasingly replaced by economic calculation and consumerism. We might become so connected that we lose our ability to truly connect with each other. We might stumble into a world in which we prioritise an easy consumer experience over jobs, wages, dignity and rights, with victims yet unseen.

Our experience of online education during lockdown demonstrated the extent to which technology can widen educational inequality if we do not pay greater attention to how it is used, and should provoke vigilance about how some big technology companies might use artificial intelligence (AI). Lucy Kellaway, a newspaper columnist turned teacher, has written powerfully about the evidence of this divide.

'What is scary is not the size of the average impact of the lockdowns, but the unfairness of its distribution,' she writes. 'If I think of the students I teach, the top quarter of the class appears no worse off after having missed so much school. They have mainly toiled diligently online through two lockdowns, and any small holes will be easily filled. Their lifetime earnings won't suffer one jot. Even for the middling students, what surprised me last time was how fast they caught up. By contrast, what surprised me in the opposite direction was the dire straits of the bottom third of the class – most of whom are from deprived backgrounds. The ones who do not speak English at home were in the worst state of all.'

The genie will not go back into the bottle. We can't prevent massive change as a result of the technology tsunami, even if we wanted to. So the challenges we face require us to befriend the machines, to learn their language and to work out what problems we can solve together. We can ask – persistently – the right questions about what technological change means, and what we want from it. The answers should not just be about efficiency, speed and cost. We should also acknowledge the victims of technological progress. The new lords of this world must find ways to share the benefits with those the computer scientist Jaron Lanier has called 'digital serfs', if they are to avoid their own King John moment of accountability. Or, as Kevin Williamson of the *Washington Post* put it, 'The new emperors should learn from the old emperors that empires fall.'

As I have argued throughout this book, we need to mind the gap. If technology does indeed entrench existing inequalities and create new ones, the challenges I have described of living together will be insurmountable. Our collective survival chances depend on us cracking this problem.

When people debate these risks, no area of technology is misunderstood more consistently than that of artificial intelligence. At one extreme we are presented with scenarios where AI gradually – and then probably suddenly – replaces us. At the other extreme we look back at our obsession with it as a peculiarly 2020s fad.

Perhaps neither scenario is really accurate. Strip away the hyperbole, and we can say with greater confidence that AI will eliminate the need for humans to perform the laborious and tedious tasks that have taken up so much of our time throughout history. It will indeed change or destroy some jobs, but it will also create new ones. It will help us anticipate trends, and prepare for the future. It will help us get better at responding to the natural disasters that the climate crisis will make part of our lives. It has already allowed us to detect diseases faster, to solve problems more quickly. It will reduce the risks associated with human error and human weakness.

Alongside these benefits, it is of course right that so much public debate should focus on the risks of AI, from the threat to industries to killer robots. We must also heed Henry Kissinger's warning that 'we must expect AI to make mistakes faster – and of greater magnitude – than humans do'. We are already seeing the ways that AI can indeed be a tool for discrimination, a source of unemployment and a means of curtailing human rights. Joy Buolamwini has, for example, shown how facial recognition software can often be used in a way that is discriminatory. She has identified racist and sexist bias in algorithms. Through her Algorithmic Justice League, she has managed to get companies such as Microsoft and Google to change their approach.

But 'AI is neither good nor evil,' Joanna Shields told me. 'It's how we use it that makes the difference. It's up to us to ensure technology is used for good. We can't just wait and see what Big Tech will do. I know because I had a front-row seat working for tech giants like Google, AOL and Facebook.'

Joanna spent years in what many would now consider the belly of the beast, working for some of the best-known Big Tech companies. She emerged with a determination to help the world understand the scale of the challenge we face. Where did we go wrong? 'Technology promised a more connected world with greater equality and social progress,' she says. 'Yet what we got was something of a free-for-all with tech giants living by the hubristic motto of "move fast and break things" without any consideration of the damage being done. It was an exciting and even an intoxicating time. But no one stopped to calculate the unintended consequences. Nobody thought of the potential misuse by malign actors. No frameworks or blueprints were built to govern emerging technologies and platforms.' At the heart of the challenge was that technology simply moved too fast for governments, which ended up playing a version of 'bash the rat', trying to manage issues that had been overtaken by the time they understood them.

Now focused on the benevolent uses of AI, Joanna argues that we don't have the luxury of failing a second time round. 'AI can either alleviate or exacerbate divisions but the stakes are exponentially higher,' she says. 'We need to pay close attention to the bias that seeps in from humans to machines. Technology propels us forward in so many ways, but it is not a surrogate for society.' This means that we cannot abdicate our responsibility to use technology safely and ethically. For Joanna and others who have emerged from the first digital revolution, there is an urgency to the need to learn from the mistakes we made and to put humanity at the heart of all we do. 'AI alone is not enough, for once an AI platform or system has made a prediction or recommendation, it requires interpretation by human expertise and capabilities.' Used wisely, technology can *augment* our capabilities and unleash the transformative potential of our creativity. 'AI may indeed boost our productivity and economic development, but no monetary value will make up for the damage if progress comes at the expense of our humanity.'

Joanna recognises that AI faces a challenge of trust, too. AI can't make meaningful progress until the public understand and believe in it. This will happen only by allowing the public to scrutinise the way AI is developed and deployed. 'AI cannot only be imposed on people as a *deus ex machina* from above with no explanation whatsoever. It needs to be discussed with the public and governments and organisations need to carefully listen to the public's feedback and concerns. Scrutiny, accountability and transparency must be the key pillars to promote digital literacy and gain public trust.' If we get that right, we have a better chance of ensuring that the AI we are building is magnanimous, not malevolent. 'The diversity of the data we use to power our AI and machine learning must represent society as a whole so that no one is left behind. Building trust takes time and you can't rush it.'

We therefore have a collective responsibility to put in place the frameworks to mitigate those risks now. We need new global standards, 'Statutes of Liberty', safeguarding freedom of expression and setting out parameters of net neutrality. Normally we would look to governments or international organisations to do this. I do not believe we have that luxury. Many Western governments have long lost anything vaguely resembling a monopoly on information and influence in this space. Others, such as China or Russia, have a completely different concept of rights and responsibilities online. And so the effort to build better safeguards will be painfully slow and inadequate. Those seeking to evade them are better resourced, faster moving and more agile.

Of course, tech bling and must-have gadgets aside, arguments over the balance between freedoms and security pre-date the digital era. Like the Magna Carta barons, we still need to understand where authority begins and ends; what issues fall under the rule of law; and how to balance the rights of individuals and communities. The Nobel Prizes of the future may be awarded to those protecting our neurological rights – our freedom to think and to reason – in the face of AI. Some governments are already taking innovative steps to protect neurological rights online. Chile has been an innovator in exploring laws to counter the way that technology can be used to reduce our liberty.

That is not to say that rules don't exist already. For example, there are existing regulations for data privacy. Article 8 of the European Convention on Human Rights says that 'Everyone has the right to respect for his private and family life, his home and his correspondence … there shall be no interference by a public authority with the exercise of this right except such as is in accordance with the law and is necessary in a democratic society in the interests of national security, public safety or the economic well-being of the country, for the prevention of disorder or crime, for the protection of health or morals, or for the protection of the rights and freedoms of others.' Article 17 of the UN's International Covenant on Civil and Political Rights says that 'No one shall be subjected to arbitrary or unlawful interference with his privacy, family, home or correspondence, nor to unlawful attacks on his honour and reputation … everyone has the right to the protection of the law against such interference or attacks.'

But we rarely have the power, the patience and the will to enforce these rules or to hold Big Tech to account. In 2018, there was a striking and symbolic moment of the transfer of power when a UK parliamentary committee tried to summon Facebook founder Mark Zuckerberg to a hearing. He declined, so they offered to go to him. He sent an underling.

If governments cannot do this alone, part of the answer is therefore to bring together coalitions from science, industry, civil society, international organisations, academia and politics to put in place the scaffolding for collaborative governance of new technology and to ensure that AI and other fast-moving technologies are a force for good. Any organisation governing or regulating AI should be multilingual and multicultural. It should be independent from government. This motivates my work with the Global Tech Panel and the Oxford Commission on AI and Good Governance. Both bring together experts from the industry with academics, policy planners and human rights activists to work on practical ways that we can make the most of technology.

A diverse coalition of civil society actors have articulated the International Principles on the Application of Human Rights to Communications Surveillance – a framework for understanding the impact of surveillance on human rights – and have outlined steps to guarantee transparency and accountability from both states and industry alike. The principles include only permitting communications surveillance by state authorities where the aim aligns with democratic norms. Any measure taken must not be applied in a manner that discriminates on the basis of race, colour, sex, language, religion, political or other opinion, national or social origin, property, birth or other status. We should know when we are being watched and why. There should be independent oversight.

In the absence of many governments and international organisations, some civil society groups are also trying to fill the gaps in how we monitor such abuses. Human Rights Watch has documented how governments and companies restrict online speech and access to information. It also investigates 'how digital surveillance tools, from hacking to facial recognition, are used to target activists, racial and ethnic minorities and workers'. It seeks to 'expose the impact of AI and other data-driven technologies on the rights of workers and people living with poverty'.

Ultimately it will be individuals who will need to lead the debate about how we shape the digital services we use rather than simply inhabit them.

This can provide the basis for a serious conversation about how society adapts to the internet in a way that preserves maximum liberty and transparency. We can't just mooch through these questions, waiting for our next dopamine hit, or get distracted by photos of cute cats.

As part of this effort, we as individuals need to take back control of our privacy. Professor Carissa Veliz has recently published *Privacy Is Power*. She told me her mission is to shake people out of their complacency over the way our data is used. 'This is not something we have to accept as the price of all that we enjoy online, it is an unacceptable breach of our rights,' she says. 'There is a myth that if you have not committed any crime, you have nothing to fear, nothing to hide – I think this is very misleading and dangerous because privacy is not a personal preference. It is not whether you like a flavour or not. It is a political concern. And the stakes are so high that our democracy depends on it.'

We now realise the extent to which our lives are being captured by our devices, and the way that data is then sold to companies that want to sell us their products, or as their products. Often we don't realise the extent to which this is happening, or choose not to care. As Carissa argues, 'When companies collect your data, it doesn't hurt, you don't feel the absence, you don't see it physically.' They are usually doing so in unobtrusive ways, on the basis that it makes our lives easier. For example, I find it hard to imagine a world in which I'm not using Google Maps and car satellite navigation. But how much of my life has been recorded somewhere as a result?

Carissa's warnings come at a time when more of us are having to share our data in order to travel, shop and work. Contact-tracing apps and the sharing of medical records will become even more prevalent as we deal with new pandemics. It has traditionally been after military or terrorist attacks that governments have eroded our civil liberties. At those moments, we have often been more willing to give up significant long-term liberties to deal with a short-term challenge. Pandemics have added a new dimension.

While there have been successful examples such as GDPR, the change in how we protect our data will ultimately come from consumers rather than government regulation. Good regulation makes sure one type of power cannot be transformed into another. Carissa cites the example of money. Money is powerful, but we have measures in place to reduce, not always effectively, the way that money can influence our democracy (such as through buying votes and buying politicians). Yet currently, data and those with data are not subject to the law in the same way. We do not need to give up data, but we do need measures to protect people from misuse of the power it gives to governments, institutions and individuals. We saw, for example, during the 2021 Hong Kong pro-democracy protests, the way that facial recognition surveillance can be manipulated, forcing protesters to adapt their tactics in order to exercise their rights to protest. Surveillance cameras in Chinese universities are able to observe which students are more inquisitive. The idea of governments using tech against their citizens is not a dystopian future; it is already happening.

People from vulnerable groups or societies understand this as second nature. As Carissa has argued, when we are told that a measure is for security reasons, we should always ask: security for whom?

I asked Dr Noran Fouad, a Blavatnik School of Government expert on protecting ourselves online, why this matters. She presented it as an issue of not just protecting our own data, but a wider obligation to society. Cyberspace is complex and interdependent: our security depends on the protections others put in place, and vice versa. She pointed to measures to increase access to the internet in less-developed regions. On the surface, these are an excellent way to spread opportunity. But many people cannot afford protection from viruses and cyber-attacks. They often live in areas where cyber laws are not enforced, or which have no treaty agreements robust enough to protect them.

What can we learn and practise as individuals to give ourselves a better chance of surviving the tech in our hands? The survival skills that this book has described – from curiosity to empathy – are a start. But we will also need to study and hone five more survival skills: the ability to work alongside technology; an understanding of how our relationship with technology changes us; critical thinking; a better radar for the threats that come from the internet; and how to be more human.

To work more effectively alongside technology, we need to learn how to approach the internet with the fascination, agility and curiosity we looked at in Chapter 2. Tabitha Goldstaub chairs the UK's Artificial Intelligence Advisory Council. I asked for her advice on what adult learners need to know to be ready for the coming changes in technology. 'I think we have to ensure that we aren't fearful about technology,' she said. 'They need to see it as man, woman and machine coming together. Like superpowers, rather than something that is going to take their jobs. The key is that we are able to do many different things at one time. We can adapt throughout the day. So what we should be doing is developing the cognitive skills that we have as humans to adapt and work with other people. It is all at our fingertips. Rather than waiting for our employer or our friends to explain technology to us, I think even just the act of going out and seeking this information is helpful.'

Mike Summers, a vice president at Dell Computers, sees that 'there is so much information available that it is almost too much, and if people aren't prepared to process the information effectively, it almost freezes them in their steps'. Used wisely, the internet can provide us with great health benefits: a sense of other people's experiences; expert health information; emotional support; community; and a chance to express ourselves, build relationships and find our tribes. It must have saved many lives and jobs during the 2020 and 2021 lockdowns.

But those benefits depend on our ability to seek them out. According to *Forbes* magazine, adaptable people experiment; they see opportunity where others see failure; they are resourceful; they think ahead; they don't claim fame; they stay current; they see systems; and they know what they stand for. That's a pretty useful guide for all of us as we seek to manage our time online and offline. It aligns with earlier survival skills we have looked at, from building crisis resilience to finding purpose and building coalitions rather than going it alone. But it won't just happen: as with the other survival skills, we need to observe, learn, practise and teach it. That way we make the technology work for us, and not the other way round. Or, as former head teacher and education pioneer Anthony Seldon puts it, we ensure that technologies such as artificial intelligence liberate rather than infantilise humanity.

Finale Doshi-Velez, a professor of computer science at Harvard, thinks we'll soon understand how to make decisions with machines. 'I'm excited about combining humans and AI to make predictions,' she says. 'Let's say your AI has an error rate of 70 per cent and your human is also only right 70 per cent of the time. Combining the two is difficult, but if you can fuse their successes, then you should be able to do better than either system alone.' The psychologist Daniel Kahneman has shown the way that 'noise' (the tendency of humans randomly to reach very different conclusions) and bias (our tendency to do so for a reason) can be reduced by combining our insights with those of an algorithm.

* * *

The second of those technology-facing survival skills is to understand more about how the technology works, and how our interaction with it changes us. Not everyone needs to learn coding. But humans need to know how machines are simplifying human thought processes, telling us how and what to think. If left unchecked, these new technologies may embed and standardise the ideologies and philosophies of their founders, hampering our ability to promote diversity and creativity, and systematically restricting the advancement of human civilisation.

We know that much of our addictive and often warped relationship with the online world is the result of companies having deliberately manipulated us. Our attention – what the industry calls our eyeballs – has become the world's most-valued commodity. We need to defend that commodity as fiercely as we would our own money and security, because it is bound up with both. As consumers this need not mean rejecting all attempts to gain our attention, but being attentive to why it is being sought and what they are doing with it. As companies or individuals in the attention business, we need to ask ourselves why we are seeking it and what we are doing with it.

We can also become savvier about how social media algorithms work on us, not for us. As analyst Adam Elkus has put it, 'The machinelike process of engaging with social media is sufficient to produce an unhappy consciousness that stems from self-loathing about one's own participation in a system that affords no special privileges to being human, blurs the distinction between human and machine, and recasts everything fed into it into networks and data. This self-loathing is matched with a powerful cynicism about a fake world filled with fake people and populated by fake beliefs and sentiments. Everything's fake, everyone's a bot, and you can't shake the nagging suspicion that you're a fake bot too.'

Sometimes this simply means learning when to switch it off. The hours we spend online are increasing rapidly: over two hours a day in the UK on social media alone. Social media is more addictive than cigarettes and alcohol. And it is of course more available to us, without the same restrictions or stigma. Studies have shown that just having a phone in your pocket reduces the quality of dinner conversations, and decreases the sense of human connection.

In trying to set those boundaries, it is interesting to observe that different platforms have different effects on our well-being. YouTube comes out net positive, particularly on access to information, countering loneliness, allowing space for expression and community building, though the platform scores low for its impact on sleep. Twitter, Facebook, Instagram and Snapchat are net negative because of their impact on anxiety, the fear of missing out and bullying, even though they prove net positive on self-expression and identity.

We can build in genuine time away from the device. And we can monitor where we are spending our screen time. List the hours online. Over a week, how much of that is helping us live happier lives? This may help us to embrace JOMO (the joy of missing out). It might free us to stop wondering what everyone else is doing and simply to be in the moment.

Third, to coexist with technology we will need to get much better at learning how to think critically. Access to all this information is not the challenge, but what to look for and how to use it. This requires us to separate fact and opinion, especially as we navigate an increasingly complex and sophisticated web of fake news, trolls, clickbait and memes. The idea that truth and reason are somehow less certain, less solid, has been given rocket boosters by the internet. Conspiracies fester and rationality flounders. As a recent

White House spokeswoman put it, 'We have alternative facts.' Fake news weaponises intolerance of difference and diversity, and creates a wall of noise and distraction. We feel unable to keep up or discern fact from fiction, overwhelmed by information, fearful of missing out on the latest celebrity scandal, tweet or cute cat. The modern snake oil salesmen are now masters of clickbait, creating just enough intrigue to coax us to want more.

As we have seen, some governments are also deliberately ignoring the facts. Denial of the impact of climate change threatens the agreements on which our children's life expectancies depend. So we must ensure that the next generation is equipped not just with knowledge, but with the critical thinking necessary to sift through the information in front of them.

Stanford University has measured what it calls 'civic online reasoning' – young people's ability to judge the credibility of the information they find online. Most students lacked the basic ability to recognise credible information or partisan junk online, or to tell sponsored content from real articles. As the team concluded, 'Many assume that because young people are fluent in social media they are equally savvy about what they find there. Our work shows the opposite.' Schools in Finland are ahead of most in trying to find new ways to combat this, developing specific curricula and assessments on how to curate and sift through online content. If in doubt, use a fact checker like FactCheck.org. For parents, we need a conscious effort to discuss and debate what our children – and we ourselves – are finding online.

As part of that education, we need to approach the internet ready to be wrong. There are browser extensions that allow you to monitor your newsfeed, and that of your friends, to gauge how much of an echo chamber you are operating in. If you seek out news outlets and sources that don't support your view, it becomes easier to anticipate change and to forecast more effectively.

Fourth, we need to learn greater caution, vigilance and self-protection, of ourselves and others. Despite the realities of current data surveillance, Carissa Veliz is optimistic that people can begin to restrain governments and companies who will come for their data. I asked her for examples of the steps we can take in our own lives. She recommended swapping Google for a more secure search engine such as DuckDuckGo; placing a piece of tape over the camera on your laptop or phone when you aren't using it; not accepting cookies; minimising your use of smart devices such as Amazon's Alexa; and using virtual private networks.

Every datapoint you keep to yourself is better, so say no to companies when asked for your data. If asked, express your interest in privacy settings. Privacy is a competitive advantage, and we will get better services if more people opt for it. Do not be predictable online: don't leave a trail of data behind you by sharing your contact details when there is no need, or subscribing to round-robin communication. Demand not only rights, but also enforcement of those rights.

I found just following Carissa's advice gave me a sense that I was reclaiming some control. Ciaran Martin, who set up the UK's first cyber defence agency, agreed. Cyber-attacks are growing twice as fast as the number of smartphones. He recommends always taking system updates, which are usually aimed at dealing with specific threats to data. Despite the hassle factor, he says we should use multi-factor authentication wherever possible, and back up our data so that if the worst happens we will lose only a copy of it. A password manager might seem a risk but it makes us less likely to use predictable or repetitive passwords. Ciaran warns of three particular warning signs that we might be under attack: messages pretending to be from tax collection offices; email addresses or URLs that don't look quite right; any promise of cash. If it sounds too good to be true or practically implausible – it probably is.

Much of this also comes down to common sense. As Ciaran says, however realistic the motorway sign telling you to close your eyes while driving, you wouldn't do it. However realistic the letter from police telling you to leave your front door open, you wouldn't do it. We have to train ourselves to be as sensible online. And we must help older people – often those most likely to be targeted – understand how to be safe.

Protecting ourselves also means that we need to be as vigilant – in fact even more so – for online bullying as we would for bullying offline.

Eighty-eight per cent of teens have seen someone be mean or cruel to another person on a social media networking site. We know that internet use is linked with anxiety and depression: rates of both have risen among young people by over two-thirds in the past twenty-five years. It is also linked to problems with body image: nine in ten young girls in the UK say they are unhappy with their bodies. And bullying: seven in ten young people in the UK have experienced cyberbullying, over a third frequently. 2021's Everyone's Invited has reminded us of the extent to which social media can be a tool for harassment and intimidation, especially of young women. Social media also drives us constantly to compare ourselves with others, which affects self-esteem. Ten million new photos an hour on Facebook provide constant reminders that other people are having a better time.

My NYU research team found huge concern among teachers about the subtle ways in which digital technology was being used not just to convince kids to eat more sugar, swear more or objectify others, but to become more violent. They even reported pre-school kids being exposed to horrific images hidden in the videos we let them watch when they need a bit of distraction. My thirteen-year-old regularly shares interesting facts from one YouTuber. Most were fascinating, but every now and then one would arrive with a conspiracy about the Rothschilds, or a controversial take on genetics. We'll need to have these conversations much earlier and in a much more candid way with the next generation than we did with our parents. Do we want four-year-olds to get sex education from teachers, as in the Dutch system? Or via Snapchat?

Finally, to have a chance at surviving this tech revolution, we can focus on learning what it means to be more human. Tabitha Goldstaub urges people to think beyond the tech gaps in their understanding of the world. 'Go to art galleries, go to the theatre, think about what makes humans human,' she says. 'And that in itself will then set you up as you start to see new bits of technology roll into your life. What makes you unique? Why would you be better than a machine at something? If you can find things where you're unique, you're more likely to flourish.'

That is why, Anthony Seldon told me, even though we may not yet fully understand the impact of AI, we must ensure that wherever these technologies evolve, we don't impose a standardised definition of intelligence as we did in the past. And we need to encourage creativity, for creative thinking and creative jobs are the most future-proof activities against automation. 'Too many national governments are competing to be the home for artificial intelligence,' he said, 'but not preparing their population for the changes AI will bring.'

Ultimately this may lead us to make peace with the machines. As neuroscientist Tara Swart put it to me, 'The illiterate of the twenty-first century won't be those who can't code, but those who don't solve problems using a combination of computational thinking with empathy, intuition and creativity.'

We must escape our fears, fuelled by decades of science fiction, that we are in some kind of neverending adversarial struggle with the machines. Instead we need to work in partnership with technology, harnessing our relative strengths. The key to living with the machines is not to be more machine but to be more human.

9

How to Be Global

I am not an Athenian, nor a Greek,
but a citizen of the world.

Socrates

Five of the world's top education pioneers struggled for space
on Richard Branson's Twister mat. I had asked an education
minister to play the part of governments, a professor to represent universities, a campaigner to take the role of parents and
a teacher to represent schools. The idea was to capture the
chaos that is global education reform, with all its competing
interests and constituencies. And to work out together – literally in this case – how to disentangle ourselves.

   In the middle of the mat, calm as always amid the chaos,
was Andreas Schleicher. This tall, thoughtful German data
scientist was representing international organisations: his day
job is director of global education at the Organisation for
Economic Co-operation and Development (OECD). As the
teachers and activists listened, he once again showed why he

is seen not just as the most influential education thinker on the planet but as a quiet yet vital rebel against entrenched thinking.

'If we're scared that human jobs will be automated, why are we still teaching kids to think like machines?' Schleicher said. 'Artificial intelligence should push us to think harder of what makes us human. If not, the world will be educating second-class robots and not first-class humans.' He worried that we are not teaching young people what they need to thrive and that students who grow up with great smartphones but poor education are at unprecedented risk. 'In the past, teaching content knowledge was the ultimate goal of education but we are no longer rewarded for what we know,' he said.

Schleicher knows what he is talking about. Every three years, he and his team have compiled the results of the Programme for International Student Attainment (PISA) test, taken by over half a million kids in more than eighty countries to assess the quality of education around the world. Governments hang on the results, terrified that they will plunge down the rankings. Many commentators have rightly questioned the focus on academic tests – in 2014, a hundred academics said that PISA was 'impoverishing our classrooms'.

Yet it turns out that Schleicher himself is its biggest critic. 'We have to change what we test if we are to change the way we educate young people,' he told me. He has been in a years-long arm wrestle with education bureaucrats to focus much more on assessing skills and creativity than maths and science. He has also tried to show the link between wellbeing and learning. Kids in the UK, Singapore and the US tend to be more anxious than others. Much better in this respect to be Finnish or Estonian.

Most important, Schleicher's years on the education front-line have led him to conclude that we must teach and assess what he calls global competence. The ability, basically, to develop the cultural antennae to thrive in such an interconnected world: 'You do not understand your own language until you learn another language. You do not understand your own identity until you understand other identities.' He is under no illusions about the obstacles to such work. Many governments, including those of the US and UK, opted out of the process to assess global competence. 'Our industrial education system is a relic of the nineteenth century,' he told me. 'It may be the young people themselves who have to change it.'

Schleicher is getting support from unexpected quarters. Chinese tech giant Jack Ma started life as a teacher. 'I learned everything in life from that,' he told me at an education conference in Paris. 'I failed my exams many times. At Alibaba I was a "chief education officer" more than a "chief executive officer".' As well as rewarding teachers more, Ma says we need a huge overhaul of exams. 'Why is it that I have to retrain university graduates when they join my company? A university degree is just a receipt for the money you spend on the tuition.'

I asked Jack what he thinks kids need to learn instead of the rote learning and memorisation that dominate current education. He was unequivocal: 'To confront failure, so that they can learn from risk. To be more creative, so that they can work in teams. To learn empathy, so that they can understand those around them. In the twentieth century, you won by caring about yourself. In the twenty-first century by caring about others.'

Like Schleicher, Ma believes the missing ingredient is that of young people learning how to interact with the world beyond their borders. 'The world of tomorrow requires people who can get out of their comfort zones. Kids need to start practising that.'

So are Andreas Schleicher and Jack Ma right? And if so, how can we help our kids to develop the vital survival skill of global competence? And how can we too be more globally agile? There are three growing reasons why this matters: automation, employability and the nature of the challenges we face.

As the first chapter set out, automation means that our kids will have to develop a very different skillset to us. As industries rise and fall faster, humans will need to move to where the opportunities are, not stay where they were.

Second big reason: employability. Like Jack Ma, business leaders consistently say that education is not providing young people with the social and emotional skills they need to solve problems, work with new information, or fit in well with a team. They say that the skills they look for now are those with which humans have an advantage over machines: adaptability, creativity and teamwork. Employers want people who can adapt to new environments.

Most important is the third reason: as we have seen, humanity faces an age of climate crisis, more mass epidemics, huge migration, new forms of warfare and the rise of rival intelligences. All of these massive challenges are global in nature. The world needs people who are able to work comfortably across national and social boundaries if we are to be ready for what lies ahead.

In an interview on the future of learning, Andreas told me that small countries are often better at developing these skills than larger ones: 'You go to the Netherlands and Germany, neighbouring countries, very similar in cultural contexts, but the Netherlands are open to the world, aware of the world; you switch on the news in the evening and it's all international news. You go to my country – Germany – and you see very little of the outside world.' So which countries fare better at the international tests? 'Singapore and Canada do consistently well. Colombia do poorly at reading, maths and science, but exceptionally well on global competence.'

How do you test for global competence? The OECD has developed assessments that check how students synthesise knowledge and make connections between ideas. So, for example, you would be rewarded for spotting the link between the climate work you have done in biology or chemistry, the political work you have covered in history and the documentary you saw on the next climate summit. Global competence tests are more likely to reward collaboration and teamwork than solo learning. Students would be encouraged to show how they can identify and understand other perspectives to their own.

That openness to other people, and to other cultures, is also an area in which girls do distinctly better than boys in every country for which we have comparative data. 'Global competence is empathy at the aggregate level,' explains Andreas. 'Empathy has a personal dimension: to what extent am I open to the person in front of me. Empathy also has an aggregate dimension: to what extent can I engage in different ways of thinking like different cultures.'

When I mentioned that my parents had a poster on the wall saying 'Never let school interfere with your education', he told me that the power of governments to dictate what people learn is dramatically diminishing. 'The high rates of dropout we see in many countries has little to do with economic reasons, it's because young people do not see school opening life opportunities for them. One of the most amazing experiences for me was when I did a school project in Japan after the tsunami there. It had almost wiped out the school system, but it created a huge openness, and schools networked, so they were studying with teachers in Australia and New Zealand,

and it transformed the system. Technology now provides you the tools not just to learn from the teacher next-door but the best possible teachers anywhere in the world. But this is where global competency comes back, because if you don't have that level of openness and willingness to engage with diversity, you are more likely to shut your doors than to open your doors.'

At this point, let's bring back in Roly Keating. Watching those crowds gather in the piazza outside the British Library, he told me of his fear that we live in an age of 'politicisation or simplification of the historical record, or myth-making of various kinds, whether it's about national identity or religious identity or politics or individuals'.

And the solution? 'To pay tribute to human contradiction and the complexity of the record. The deeper you go into the record, the less mythologised and simple it gets. So our message to young people, to all of us, has to be that when you think you've got the answer, ask again. Because there's another layer to it. Please don't imagine you can read one book and that will somehow give you the answers or the knowledge that you need. Read the next one. The best critique of the book you just read is the book you're about to read.'

I asked Roly where that understanding came from. 'I think most human beings are born with a version of it. And maybe it can get suppressed. My hunch is it's the natural state. But whether through economics or underfunded education, or just sheer desperation of a situation, people aren't able to let it – maybe – grow. I was lucky to grow up in a household where it felt natural to do that.'

The new dividing lines will be between those who can embrace this readiness to question themselves, to embrace new ideas and cultures, and those who feel left behind by it. Those dividing lines will no longer necessarily be about wealth, or nationality, but whether we are open or closed to the world.

What skills should we focus on if we are not to end up on the wrong side of that line, to be sufficiently global? In the last quarter of a century there has been a growing effort to tackle this question. The Brookings Institution has developed the Learning Metrics Task Force, the most comprehensive list of the different frameworks that exist for learning. It took over three hundred lists of key skills and boiled them down to six: critical thinking and problem solving; collaboration and influence, especially with one's peers and within diverse groups working across national borders; mental agility to adjust for changes and unknowns and to upskill quickly and as needed for career needs; entrepreneurship to identify and act on new ideas; effective communication; the ability to access and analyse information; and curiosity. Rebecca Winthrop, who led the project, told me that it would have been better to keep the learning goals even simpler: 'Life, work and citizenship'.

As part of its 2030 education framework, the OECD has set out more detail on global competences, and worked with the Asia Society on ways to make them practical. They conclude that the key is 'the capacity to analyse global issues critically and from multiple perspectives, to understand how differences affect perceptions, judgments, and ideas of self and others, and to engage in open, appropriate and effective interactions with others from different backgrounds on the basis of a shared respect for human dignity'. This includes:

- knowledge and understanding of global issues and other cultures;
- communicating effectively with people from other cultures and countries, analysing and thinking critically to scrutinise and gain information, and adjusting thoughts and behaviours to fit new situations;

- attitudes of openness towards people from other cultures, with respect for cultural difference, global-mindedness, and personal responsibility for one's actions.

If you are lucky, your kids will be at a school that is already experimenting with teaching in this way. The International Baccalaureate (IB) has a curriculum that helps learners to be 'inquirers, knowledgeable, thinkers, communicators, principled, open-minded, caring, risk-takers, balanced, reflective'. Carolyn Adams, the IB's strategy director, told me that this focus was the reason why the IB has thrived in the way that it has, thanks to 'the ambition not just to say here's an international curriculum that meets your practical needs, but to say this is something which is going to change the world and make the world a better place'. An increasing number of countries are working it out. In just a few years, Singapore has placed itself in the vanguard of education reform. At the heart of that is the way that it has reformed its curriculum to help students become confident and independent thinkers, self-directed learners and concerned citizens. Aware of growing anxiety and stress levels among young people, Singapore has introduced more modules with no connection to exams and assessment.

Building on this progress, Harvard education gurus Fernando Reimers and Connie Chung have even developed a curriculum for global citizenship. Fernando told me that we are a long way from countries adopting such a change, 'but I don't doubt for a second that you can build coalitions with various individuals and groups and networks, interested in global citizenship, and that they will do it.'

I asked Connie if there was a danger that the effort increased inequality, because it gave another advantage to those who could afford to send their kids to schools in the vanguard. She agreed that we were at a moment of peril. 'The danger is that those who can't access these global skills will fall further and further behind,' she said. 'We have this underlying assumption that if we equip individual children with these skills, inequality will disappear. But looking at the distribution of wealth, I feel governments are losing power to corporations. How do you educate in a way that helps young people understand these complex global dynamics? That myth of if you study hard you're going to do well is not going to be true for a lot of our kids. If your end goal is a good future for everybody, then education is probably the weakest link.' But Connie also argued that this is an area in which some countries can play catch up, because if countries that are doing low on the tests don't adapt quickly, others will close the gap. 'Technology can be incredibly useful for someone like a Syrian refugee who can't access schooling in some other way,' she said. 'If they can access through the internet a well-designed, thoughtful class, that's a really quick way to increase the opportunities that are available to somebody who otherwise will not have them.'

BBC education correspondent Sean Coughlan rightly cautioned me that we must be careful that the idea of 'global competency' is not a Western export. Schleicher agrees. When introducing the idea of global competences, he deliberately avoided the label of global citizenship because of the connotations of belonging to a particular nation state.

The challenge for the next phase of education reform is to find ways to measure these skills, so that governments, educators, learners and parents are more incentivised to take them seriously. The OECD does this through a combination of a cognitive assessment and a background questionnaire. I've also worked with experts at Cambridge University and

UNICEF on a global learning passport, which could allow young people – especially migrants and refugees – to carry qualifications between national systems. The current system places those who move between countries at a significant disadvantage. The learning passport enables – online or offline – continuous access to quality education. Powered by Microsoft, it is easy for countries to adopt, or to use alongside national digital learning platforms.

The greatest challenge is not only to ensure that populations at risk, largely refugees and forced migrants, can continue to participate in beneficial education programmes along their journey, but also that these offer them the right syllabus content to follow at the right time, thereby minimising educational disruption. The passport could therefore address the problems linked to educational mobility in emergencies, by providing refugees, migrants and displaced children with a digital tracking mechanism of the subjects they have already studied, as well as a roadmap of those that should follow. To succeed, we will need to identify common ground between education models across the world, so that these children can continue their education with as few disruptions as possible. There must be common elements to maths or science curricula, for example, or to language proficiency. Perhaps this work could not only address a significant bottleneck in education for those on the move but become a building block in the development of global learning goals and even a global core curriculum.

Of course, a key part of the education system is not just how we decide what to learn but how we assess what we have learned. As part of my work at NYU, I also explored the potential for a credible global assessment of adaptability for an age of massive human migration. We considered the most important features of global agility. One striking conclusion we came to was that often those who had experienced more disrupted lives were those best able to demonstrate these skills.

For example, young refugees often had a highly developed radar for opportunities. They were used to scanning the environment for possibilities and risks. In an educational context, we should be helping young people to learn these horizon-scanning skills, something that can partly be achieved by studying the language, history and culture of other societies, but which must also be approached through the understanding of great cross-border themes, challenges and crises: migration, culture and the arts, supply chains, pandemics, global brands and technologies, environmental issues, global institutions. The most globally agile students were excited about discovering more about the world. This was not – as my generation might sneer – about Instagramming suitably exotic experiences or any other self-interested motivation to posture as worldly, but about meeting, interacting with and learning from different people and cultures. Without it, genuine global agility is impossible.

We also found that people who have made transitions between cultures also tended to be able to think more critically about the information around them, and not take it at face value. Fake news and the internet have shown many academically smart people that we must learn how not to be dumb, or how not to be reflexive. Globally agile humans are able to sift and combine information from different sources: books, friends, mentors, traditional and digital media. This allows them to develop greater perspective. They are more likely to see different sides to an argument. They are less likely to fall into the trap of reaching emphatic opinions without going through the substance of the argument.

We found that global agility also requires digital savvy. Not an ability to code. But more a sense of how to use different platforms effectively and responsibly, and to navigate between them. Young people who struggled to adapt were often those more likely to be propelled by social media towards getting stuck in a gang or an echo chamber. Again we found that refugee communities, frequently those who had not grown up with constant access to social media, were better able to use the access to the internet they had to establish more meaningful connections, for example with distant family, and as the key to accessing essential information. The most adaptable humans of the future will be those who have the flexibility of thinking and doing to be ready for unexpected changes. Developing this skill reduces the culture shock we all experience when encountering new places or people. It increases our resilience. But trying to respect different people's approach to life does not mean we have to agree with them, or that we should pretend the differences are not there.

Our worldviews are shaped by where we come from, and we have multiple, overlapping identities: a child can be a girl, European, tennis player, Muslim and Justin Bieber fan. The key is that she can understand how these different tribes have shaped her life, and therefore apply the same thinking when she meets people from other tribes. Our individual cultural affiliations are unique, fluid and dynamic. Our values are not uniquely special.

That does not mean that we all become, in Theresa May's regrettable phrase, citizens of nowhere. The more globally agile individuals we interviewed recognised that their cultural identity was neither fixed nor superior; they saw their self as something they could continue to inform, value and nurture. They could connect the global and the local. We need to know what baggage we carry, and why. The answer to the twenty-first century is not to let the friction of globalisation grind away our differences or let us lose a sense of where we are from. We often need to make the connection between local and global – for example, understanding the risks of climate change by studying floods in our own neighbourhood. We will be better able to tackle those global challenges if we retain an ability to root them in experiences that people understand.

The more globally agile young people we interviewed also had stronger communication skills: they were more likely to be able to interact respectfully and flexibly with other cultures. Future assessments of young people will need to evaluate whether they can read others' approaches to communication and adapt, building bridges to new people and cultures, especially those that initially appear most different. It is about listening as much as talking. And at its most effective, it also means resolving or at least reducing conflict.

The most effective global operators in our NYU study had exceptional antennae. Can you zoom out and see a person or situation in 360 degrees? This means being able to listen humbly and observe, including body language. At its best it also requires a good dose of empathy: as we considered in the last chapter, can you get in the other person's shoes, and understand their motivations? The most globally agile people care about the world around them, and think that they can make it better. They can see that people from other places share the same basic rights to food, shelter, work, education, happiness, dignity.

Helen Clark was prime minister of New Zealand for nine years. As one of the few women in politics at that time, she had to overcome prejudice and systematic discrimination. She then ran the UN Development Programme, taking that ability to break glass ceilings to the global level. I asked her how she found the courage to continue the exhausting work of leading the world's humanitarian effort, in the face of such challenges.

'It came initially from the experience of being a New Zealander,' she said. 'You feel like an outsider. And you learn that there are no doors so strong you couldn't break through them. My history teacher taught us to think differently. I remember her teaching us about the Vietnam war. Understanding the catastrophic negative impact that our actions can have opened up my mind, forced me to leave my bubble, to see things from other perspectives. That is much harder than it sounds. I was able to take that willingness to understand where other people are coming from into a political career.'

What does this look and sound like when it comes together? It sounds like fourteen-year-old Rania, from Aleppo in Syria, now living in Jordan's vast Za'atari refugee camp.

'In history, we learned about the British Empire in India,' she said. 'I was curious about the people involved, not just the battles and the business. I asked an Indian man who worked at the camp to show me photos and tell me what life was like in his country. He talked about the caste system there. He said that we are lucky not to have it in Syria. I wondered whether we do have a caste system of sorts.

'There is a boy in the camp who arrived from Iraq a few years ago, so I asked him about life there too. He told me another boy was bullying him for his accent, so I got some friends together to stop that.

'At breakfast I saw that our tea is made in India, but that it says "English breakfast tea" on the label. I looked online at what life is like for workers in the tea plantations. I want to persuade people to buy tea from places where they are treated kindly.'

SEVEN WAYS TO BOOST YOUR GLOBAL SURVIVABILITY

Becoming global is a skill that you never master. It requires a lifelong effort.

The good news is that young people, particularly digital natives, are already honing these skills, attitudes and values to a far greater extent than we realise. Look at the diversity of what children watch on YouTube or Netflix.

Here are eight activities I have used with my students that we can use to help us be more globally agile. A health warning: more than other survival skills in this book, global agility requires engaging with controversial issues. That's part of the point.

1. **360-degree analysis.** Observe a famous individual or someone new we meet. Imagine being a reporter on not just what they said, but why they said it. How and why did their background, nationality or gender influence their worldview?

2. **Role play.** Pick an international story from the news in which disagreement seems entrenched. Try to explain it from the perspectives of the two parties. How would we mediate between these competing positions? What baggage do they bring? How is the story being reported in different countries and by different news outlets, and why? Is it possible to hear from people in their own voices?

3. **Travel guide.** Discuss as a family the next holiday. What are the key bits of information to learn about the country we are visiting? What would be the best books to read or films to watch to understand how people there see the world? What are the most

important words or phrases to learn? Does that list include writers or creators from the country, not just those observing it?

4. **Hone the antennae.** Seek out the digital equivalent of a pen pal from another culture. Work together on a school or life project. Global Cities has a digital exchange programme (Global Scholar) through which kids can work together in e-classrooms across the world.

5. **Be methodical about building a tribe.** When joining a new club, school or community, work together to identify the key people to get to know: who are our network starters? What are our points of connection or difference?

6. **Know our baggage.** Discuss values as a family and as individuals. Where did these come from? How have they changed over time and generations? And why? Article 1 of the Universal Declaration of Human Rights says that: 'All human beings are born free and equal in dignity and rights. They are endowed with reason and conscience and should act towards one another in a spirit of brotherhood.' In indigenous African culture, *ubuntu* means that a person is a person because of others. Do we agree? How do others see our values?

7. **Identify the challenges we really care about.** 'Don't ask kids what they want to learn when they grow up, but what problems they want to solve,' says Jaime Casap of Google. Once we have thought about what motivates us, discuss *how* we can get involved. What information do we need? Who are our allies (and opponents)? What is the initiative we can take at school or at home?

I asked former SAS commander Sir Graeme Lamb what we should be teaching the next generation. I had expected something on courage and values. Instead Graeme reflected that 'Education today shouldn't be bothering with the three Rs of the empire. The skillset that everyone needs is to understand cross-cultural communication, body language, tone. To stand up and talk and sit down and listen. These are truly life-giving skills.'

As the late Ken Robinson argued, education should equip people to live together in tolerant and culturally diverse societies, 'to understand the world around them and the talents within them so that they can become fulfilled individuals and active, compassionate citizens, able to build lives that have meaning and purpose in an unpredictable future'. His TED talk on these ideas is one of the most successful ever: people know that there is a vital message here. Ken told me that the question he was asked most was 'Where can we go for ideas on how to get that understanding?'

I hope this book helps. Future humans should be equipped with the vision, judgement and patience to be more global. We should teach students how to be citizens of everywhere, active in creating a more equal, just, peaceful and sustainable world. That takes us to the final survival skill, and the one that we won't know within our lifetimes whether we have mastered.

10

How to Be a Good Ancestor

A society grows great when old men plant trees in
whose shade they know they shall never sit.

Greek proverb

After the Fukushima nuclear crisis in 2011, a group of two
hundred Japanese pensioners volunteered to face the dangers
of radiation instead of the young. The cancer they could
develop from the radiation might take twenty to thirty years
to develop, meaning they would no longer be alive to experience it. Yasuteru Yamada, the seventy-two-year-old who
organised the retired engineers, teachers and cooks into the
Skilled Veteran Corps to help with the disaster, told the BBC
that their decision was 'not brave, but logical. The question is
whether you step forward, or you stay behind and watch.'

Like the Fukushima pensioners, our ancestors had a much
stronger sense of the circle of life, the passing of the seasons
and years. The rituals and rites of passage were hardwired
into the social calendar, and were often the glue that held

communities together. Stories were preserved, embellished, cherished, shared. Look at the early chapters of the Old Testament, which are full of lengthy accounts of family histories, or the oral traditions of passing on family stories by the fire. My father once said to me that one of the hardest things about losing parents is the realisation that you are now the story bearer. Perhaps this is why so many in the second half of their lives become so obsessed with tracing family history.

Much of the hunt for information about our more immediate ancestors starts with digging back into official records, piecing together the story of who was where when. Inevitably, therefore, it becomes an account of key dates: births, marriages, deaths. And one of locations: the moves between addresses, a chance finding of a photo or postcard from a faraway place that opens up a new mystery. We find ourselves wanting to walk where they walked, to handle objects that they handled.

There is real adventure in these searches. While preparing for my Beirut posting in 2010, I spent three weeks in Oxford, diving into accounts of Lebanon's tumultuous recent history. One aimless afternoon I decided to drop by Rhodes House. We knew that the library held some of my grandfather's papers, left by my grandmother, but we didn't know exactly what. He had been an education officer in Nigeria between the two world wars.

The Oxford library staff, champions of curiosity, love this sort of request. But the librarian came back to me with an apologetic shrug. They held just a single piece of paper, double-sided, with a list of the towns in which my grandfather had served. It looked as though he had at some point thought it worth typing these out, and then got distracted.

I ran my eye over the list, trying to feel some connection, but it was hard to get excited. Disappointed, I asked the librarian to have one more try. Could he be listed under his initials, 'AA', rather than his first name, Anthony?

She typed in the search and looked up, startled. They had twenty-two boxes of his letters to my grandmother, from 1924 to 1965, covering their courtship, engagement, marriage in 1929, and life in Nigeria and the UK. The librarian disappeared into a basement and came back with the first box, full of yellowing handwritten letters wrapped in string. I dropped everything and plunged into their world.

These were letters of extraordinary literary breadth and poetic power. They gave me an insight into life in the 1920s, with all its colour, music, knowing innocence and sense of hope. A period when a brush on the cheek could take six months of courtship to achieve, but could sustain them through months apart. A life without modern communications, when the daily writing of letters, and the agonising waits for the post, created a more intense personal connection than any tweet, text, email or Skype. And, most of all, an insight into an exceptional musician, writer, humorist, lover, adventurer, ancestor.

I was torn between fascination and a sense that the letters were too intimate to be read. Some undoubtedly were – I set aside the most personal. But, on balance, I felt that the grandfather of these letters, though modest about their contents, would not have begrudged me this insight into a man we did not know well. The man who would write in every letter of 'the duty of rememberings' left his descendants the means to share in that duty. As well as revealing a huge passion for life, and a stupendous love affair, these letters were above all about family, friends and faith. Thanks to wisdom and a bit of good fortune, they had been preserved through eighty years of travel, and had outlived their author and his muse. Tied together with knots I no longer know how to tie, these are letters that I could never have written.

My grandfather called the letters their 'indestructible links'. And in finding inspiration, echoes of evocative memory, surprises, joy and sadness in them, I felt an indestructible link to that recent past. This led me to another adventure of memory, researching my great-grandfather Gus Smith's role in the first flight along the Nile in 1914, on the hydroplane he had built. The exploration became a BBC documentary and a four-generation family centenary commemoration in Cairo.

Many of us have experienced similar exhilaration at unveiling layers of the past. One of my hopes is that advances in DNA testing will enable more people to dig back even further, following the trails through the centuries as our ancestors moved across continents. If a billionaire wanted to do one thing to help us understand how to live together, giving everyone the ability to see so far into their past this way would help us understand our shared history: we are all migrants.

It is becoming easier to pick up these trails, the fragments of memory, the indestructible links. What is harder, beyond the apocryphal tales of distant relatives, is to preserve what these characters stood for, their values. In our family narratives, we form a sense of the recent ancestors who have done most to shape us. But despite all the search-engine-propelled research, I suspect we know less about our great-grandparents than they did about theirs. Our sense of community and of calendar has been bent into a different shape by several centuries of urbanisation and decades of globalisation. Netflix and decent central heating have replaced the campfire. Yet part of honouring and remembering our ancestors is to protect and pass on the best of what they left us, including the best of those values.

Somewhere along the way, did we forget what it means to be a good ancestor?

While in Lebanon and since, I became fascinated by the work of a psychotherapist, Alexandra Asseily. At its core, she told me, her outlook is based on a simple idea: that the role of our ancestors in conflicts affects us psychologically, influences our relationships with family and friends, and contributes to our propensity to participate in the next wave of strife, and (or sometimes not) to pass it on to the next generation.

If she is right, we bear a huge responsibility for whether, through our beliefs and behaviour, we transmit these traumas and grievances to our children, an inheritance that has far more potential to shape their lives than the contents of a will. Similarly, we can see ourselves as receivers of inherited patterns and traumas, echoed from conflicts rooted before our time. Alexandra believes that it is possible, as individuals and at a group level, to address these deep-set traumas and intolerances, and to make it easier for communities to reconcile.

For Alexandra, the key questions to ask ourselves are therefore: 'How do we become good ancestors and refrain from passing on trauma or negative beliefs to future generations? How do we stop being the prisoners and the puppets of the stinging memories of strife that we can still feel today as though we ourselves were present at that first event? How do we clean up our "ancestral arteries" so that our children are free to act in the now, free from the blocks which echo from the past and clog up our todays and our tomorrows, and are able to receive the collective talents and gifts instead? My purpose is to do my best to become author of my own narratives, and to help others become true storytellers, free to tell their own authentic stories.'

I think the best way to start to reconcile with the past and the future is by reflecting on two more challenging, very personal and fundamental questions.

What did I inherit in terms of family values and history that I MUST pass on?

What did I inherit that I MUST NOT pass on?

These are the simplest questions. And yet the hardest to tackle. Most people spend a lifetime figuring out the answers. Practically the process can also involve probing our family memories and stories, what Alexandra calls a 'great architectural dig'. But as with archaeology, we may find things that are buried, which 'require our knowing, acknowledging and understanding, as well as a "carbon dating" in order to reveal how, and why, they were hidden and buried'. In all of our families there may be ghost stories, says Alexandra, the difficult or troubled individual, the moment when someone was rejected or driven away. 'As we bring these old elements to the light, they can be appraised with love and trust, rather than by judgement or fear. It becomes possible to release old and ancient grievances.'

By doing this, we can find ways to ensure that our personal, family or community's history informs us but does not control us. We can discover the points in our family's past where the scars never properly healed. And we can give our descendants a better chance of healing or of moving beyond them. And so, in the answers to those two questions about inheritance and legacy, lie real secrets to survival, and the key to being a good ancestor ourselves.

This work goes beyond our immediate family stories. We have looked at the survival skill of taking on unfairness and injustice when we see it. Doing so even in small ways can help to create the reforms and changes that will boost our collective odds of survival. But being a good ancestor also means confronting more systemic, underlying injustice, not just the visible examples of it. It is not enough for us to tackle the injustices we can see now. We should anticipate those that our descendants will hold against us. Perhaps as well as considering what we would like in our eulogy, we can consider what we want our distant descendants to think of us. Will they venerate our statues, or will they tear them down?

The toolkit for tackling these deeper and more structural challenges is similar to that of becoming an activist: we find a cause, understand why it matters, build unusual coalitions, prepare with patience and humility, and identify the moments when change can happen. By leading the non-violent salt march against Britain's monopoly, Gandhi showed the world not just the visible injustice of the violent oppression of freedom of assembly, but the unfairness of an economic system in which Indians were denied by the British the right to their own resources. It exposed the fundamental flaw at the heart of colonialism, not just the oppression that protected the system.

We may not be able to lead non-violent marches in the face of oppression. But we can do much more to improve our antennae for detecting and understanding the underlying injustices still around us, and we can be moved to do something about them. Three of those injustices that good ancestors would try not to pass on are inherited inequality, inherited climate crisis and inherited conflict.

The UN's Sustainable Development Goals, covering education, health and humanitarian priorities, are a good place to start. Helen Clark, many people's candidate for UN secretary-general in 2015, told me that if only one international goal was adopted, it should be gender equality – empowering women has a huge impact on all the other objectives. She is too modest to point out that she was the frontrunner for UN secretary-general until fifteen male diplomats went into a dark room and voted for a man.

Meanwhile, the Black Lives Matter movement in 2020 has shattered any illusions that we are moving fast enough towards genuine racial equality. And, thirty years after homosexuality was declassified as a disease, it remains illegal to be gay in seventy countries, covering a third of the global population. It is still punishable by death in twelve countries. History hasn't ended.

For much of the pandemic of 2020–21, we found ourselves pushed towards being 'hyperlocal', more focused on our immediate communities. But it also made us 'hyperglobal': with everyone more aware of the interdependencies between communities. I think we will look back on the pandemic and see it as a moment of cultural and individual awakening. It has exposed huge flaws in our leaders. It is why in this book, I make the case that we must do a better job of teaching the skills of empathy, emotional intelligence, activism and curiosity not just to our young people, but to all of us. We all need to work on developing those skills. We come to education thinking that our role is to pass on the best of what we've learned. Yet, as part of the work of being a good ancestor, it is also about passing on some of the best of what we *didn't* learn.

To help us to challenge both ourselves and systemic injustice, perhaps the curriculum of the future could therefore teach uncertainty, dissidence, scepticism, curiosity, ethics and solidarity.

Activism on these massive underlying inequalities is of course even harder than activism on those that are more visible. It starts as ever with doing that work of seeing and understanding them, of removing our filter. We can divide the challenges we observe into those we can live with; those we can influence; and those we can actually do something about. There are practical exercises that help. Here's one. Write down the three biggest systemic advantages you have had, and how they have changed your prospects at crucial moments. It might have been the right school, the subtle advantage of gender or race at a job interview, or a word in the right ear from part of your inherited network. Be really honest, setting aside the story you might choose to tell yourself or that is narrated for you.

Then imagine the experience at those crucial moments of someone who was denied those advantages. What will you do now to even the playing field?

For me that process exposed the imbalance in gender equality at my place of work, the UK Foreign Office. Of the eleven fast-streamers I started with as a graduate, only two were women and all were white. We took it for granted that the behaviours and values of a predominantly white male group were those that helped you to get promoted. And by and large they did: the Foreign Office traditionally loses women before they get to board level or senior ambassadorships. And I arrived only a few decades after the first women to be allowed to remain diplomats after they got married.

In 2016, I led a review of the Foreign and Commonwealth Office, which highlighted these challenges. The then Head of the Diplomatic Service, Simon McDonald, made it a centrepiece of his leadership to take on these structural inequalities. As of the summer of 2021, there were women ambassadors in Paris, Berlin, Moscow, New York and Washington. My foundation also helped set up the Diplowomen initiative, which connects young women in diplomacy with mentors, and offers extra support to senior women diplomats. Better late than never.

This collective effort to become better ancestors is just as urgent in our actions on the climate crisis. There is no better example of our short-term needs and interests having harmed those of our descendants. We know that failure to reduce our greenhouse gas emissions will create water and food shortages, natural disasters, refugee crises, flooding and the mass extinction of plants and animals. Archbishop Desmond Tutu has described the reduction of our carbon footprint as 'not just a technical scientific necessity; it has also emerged as the human rights challenge of our time ... the most devastating effects of climate change – deadly storms, heat waves, droughts, rising food prices and the advent of climate refugees – are being visited on the world's poor. It is a deep injustice.'

But the rhetoric from world leaders – airy words in airless rooms – has not been matched by any genuine seriousness of intent. Austerity, insecurity, impatience and the decline of nation states have all served to distract our minds from rethinking our energy needs. Our collective actions to date have barely scratched the surface of the response that is necessary. I've been struck by how many ecologists have already said that we are too late to manage, let alone turn back, the devastating environmental consequences of the Industrial Age. And ahead of us we will have to manage the end of the fossil-fuel era without triggering the kind of major conflict that energy transitions in the past have caused.

We have seen the emergence of an overdue and essential understanding that our fate is deeply connected with the ecology of the planet. 'You never change things by fighting the existing reality. To change something, build a new model that makes the existing model obsolete,' argued the American engineer and futurist R. Buckminster Fuller. Seven countries have already legally recognised the rights of nature. New Zealand has granted the Whanganui river the same rights as a human being, as has Colombia the Atrato river. In Bangladesh, the supreme court proclaimed all its country's rivers to be alive and entitled to legal rights. But, however many climate accords and summits, we cannot wait for national governments and international organisations to lead this work. As individuals we must be part of the effort in a much more assertive way.

Climate activism can seem overwhelming, and our actions insignificant. But a group of activists, UN experts and pioneers have been identifying small ways that we can all contribute to our collective survival chances. They looked especially at those actions that have the greatest impact, are easy to do, and have a multiplier effect. If we all do those that they highlight, we can keep to the Paris Agreement's 1.5 degree limit to global temperature rises that we need to survive. In terms of assessing such activism, we have found a cause, found a tribe and now we are trying to enlist others to join it in simple and practical ways. The key is to break down this daunting challenge into manageable chunks. You could start by switching to green energy tariffs and investment/pension funds, eating more plants and walking more.

You have most work to do if, like me, you are in the group of people who shop at supermarkets and malls, own a car, rent or own a multi-room home with electrical appliances, central heating and electricity, buy new clothes, go on holiday and fly by plane. But the exercise of thinking about this has challenged me to take greater personal responsibility by swapping (to green tariffs and pensions), stopping (for example, flying or driving less, eating less meat) and supporting (for example, sharing my actions and asking others to follow).

In looking at how we can be kinder to ourselves, I made the case for a plant-based diet on happiness and health grounds. More importantly, such a diet boosts our collective survival prospects. Food production accounts for about a quarter of greenhouse gas emissions, more than all the trains, planes and cars on the planet. While it doesn't put it in these terms, the UN's Intergovernmental Panel on Climate Change warns that animal farts and burps pump methane – a greenhouse gas that is many times more potent than carbon dioxide – into the atmosphere while the agricultural land on which the animals graze has replaced the trees that would have removed carbon from the air. Scientists working on a two-year programme to find the best diet for the planet concluded that the only way we will be able to feed all ten billion of us in 2050 is if Europe and North America cut back massively on red meat and East Asia on fish. We have never before attempted to change the global food system at this speed. But we must do so to survive.

Alongside these ways in which we can become better ancestors ourselves, one of the most exciting areas of work will be in finding ways to deal with historical trauma. This is where collective psychology intersects with peacemaking and social media, and it could begin to help us, in dealing with inherited conflict, to start to heal the wounds of history.

We all carry historical trauma, even if we cannot see or comprehend it. This is perhaps most obvious in communities that have recently emerged from conflict or remain beset by it – Northern Ireland, Rwanda, the Balkans, Israel/Palestine. It was certainly the case in Lebanon, where the history books stopped when the civil war started in 1975 and a whole generation found ways to avoid having to confront or discuss the war. As the novelist William Faulkner wrote in *Requiem for a Nun*, 'The past is never dead. It's not even past. All of us labour in webs spun long before we were born, webs of heredity and environment, of desire and consequence, of history and eternity.'

Epigenetics is the study of the effect of the environment on our genes, what most of us might know as the nature/nurture debate. Are we shaped by our genetic inheritance or our experiences? Geneticists now have proof that the behaviour of our genes can be altered by experience – and can be passed on to future generations. This finding may transform our understanding of inheritance and evolution.

Experts used to think that more than anything else DNA dictated what would happen in our lives. But the science is shifting towards the idea that life experience, stress and trauma can change the expression of our genes. Our kids don't have the same genes as we do. The grandchildren of Holocaust survivors have altered stress responses because of the experiences that those survivors had either when they had a child in the womb or around the time of conception, or even before.

Neuroscientist Tara Swart has argued that there is a practical step that we can take to understand the impact of our environment on our values. Look at the five people you spend most time with, and write down their worst and best characteristics. Then reflect on whether you can see those in yourself. 'In neuroscience everything is about raising non-conscious to conscious,' says Swart. 'We know from the neuroscience that in the Israel/Palestine talks, if before speaking the negotiators made a list of ten people that they could turn to if they really needed to then they were more likely to negotiate more favourably with the other side. Just remembering that there are people there for you actually makes you more open to collaborating with someone that you see as potentially an enemy or an adversary.'

This is where science meets the politics of building a better society. In peacemaking or rebuilding post-conflict communities, we are often trying to deal with the legacy of previous trauma. Former UN envoy Bernardino Leon described much of his work in Libya to me as a form of therapy. These conflicts might have taken place long before those involved in the peace process were born. As a result, there is space in diplomacy for understanding the collective psychology of a nation and trying to be a therapist. The words you use, the tone of connection, the people you bring together, sometimes it all feels like an act of collective therapy.

We should not underestimate the extent to which that collective therapy will be needed. Many commentators in the 1910s, on the eve of a devastating conflict, believed that a massive war was impossible – the interconnected webs of trade and finance would prevent it. We should not make the mistake of complacency again. In modern times, as the US economic lead narrows, the risk of conflict has increased, as it did in the past when power shifted significantly. The power vacuum in the Dark Ages in Europe created space for religious fundamentalists, pirates and Vikings. Communities fell back on protecting themselves in smaller and smaller units. The historian Niall Ferguson has written of his fears of 'a new Dark Age of waning empires and religious fanaticism' and of 'economic stagnation and a retreat by civilization into a few fortified enclaves'. American journalist Roger Cohen has called this 'the Great Unravelling', a time of aggression, break-up, weakness and disorientation. Harvard professor Stephen Walt has predicted an increase in the number of conflicts.

We can also begin to understand the ability to use social media not just to inform or engage, but to influence communities, even nations. Look at the way that Donald Trump was able to use Twitter to mobilise his base. Maybe we can combine that ability to reach and influence populations with the collective therapy of peacemaking and diplomacy and the growing understanding of the way we carry wounds from the past. It might be our best hope of beginning to heal those historical grievances that will otherwise sustain so many of the conflicts of the twenty-first century.

Most politicians have distrusted the mob. Online mobs can at times be as vicious and easily manipulated as their offline predecessors. But technology has the potential to change power for the better by distributing it more fairly. Fewer people are being killed in conflict because the more that people are able to determine their own fate, the more peaceful they become. Our desire to network and connect is not just a fad or blip, a short-term response to a surge in connectivity. Instead, networking is intuitive, a natural way to order the world. Maybe we should learn to believe more in the wisdom of crowds after all. That might be the key to healing the wounds of history, and to becoming better ancestors.

We can aspire not to leave our descendants with inherited divisions or a battered planet. This seems daunting. But as with other survival skills, there are small changes that we can observe, practise, learn and teach that can help us to start.

One is to get better at saying sorry. Gareth Evans, former Australian foreign minister, once told me that if politicians learned to say sorry we could make rapid progress on the most seemingly intractable conflicts. The same applies to our arguments over Brexit, family feuds or what to do with the remaining council budget. In 2010 I worked with David Cameron on his response to the Saville Inquiry into Bloody Sunday, and his apology was so powerful because it was authentic and sincere – a defining moment early in his admin- istration, when he moved from being the leader of the largest party to being the prime minister. The apology recognised the context in which the British Army opening fire on Republican protesters had occurred but did not blind itself to the hurt caused. Cameron dictated most of the apology himself, and thought hard about how it would be received not just among the UK military and his own constituencies, but on the streets of Derry. Sorry seems to be the hardest word, but it is some- times the best one.

At moments in history, nations have found ways to adopt an atonement strategy, of seeking collective forgiveness.

Germany's chancellor Konrad Adenauer led this effort to atone for the Holocaust in his country's dealings with Israel after the Second World War. Only days after taking office in 1949, he set out what have subsequently become the core elements of national atonement: a verbal acknowledgement of moral responsibility for the wrongdoing; a public expression of remorse and reconciliation and the offer of restitutive actions including financial, legal or political measures: 'In so far as it is possible in the aftermath of the annihilation of millions of people beyond retrieval, the German people are willing to make good the injustice committed against the Jews in the German name by a criminal regime. We consider restitution [*Wiedergutmachung*] as our duty. The Federal Government is committed to initiating appropriate action.' As international relations scholar Kathrin Bachleitner explained to me, 'These practices are so intertwined with the German postwar case that the terms "atonement" and "reconciliation" are often described as "*Vergangenheitsbewältigung*" ("Coming to terms with the past") and "*Wiedergutmachung*" ("Making good again").' But she argues that the real reason the German model should be observed is that it was not top-down, but owned by individual Germans. She sees this as a model for other national forgiveness efforts. 'We need to bring individuals back into these political processes,' Kathrin said. 'Not as passive and manipulated victims but as active agents of peace. People are not only those most affected by war but also the agents of its legacy, for better or worse.'

Perhaps there are elements of the German atonement model that can help us to say sorry. Acknowledgement of our share of responsibility. Expressions of remorse and reconciliation. Coming to terms with the past. Making good again.

We can also practise, like the process of digging into and reconciling ourselves with our own family's history, one of the hardest survival skills: forgiveness. Before becoming prime minister, Gordon Brown wrote a set of profiles of eight individuals whose courage he admired. I asked him to define their common traits, expecting it to be about resilience or persistence. Instead he spoke of their pursuit of moral purpose. It was not, he said, that they were never fearful, nor (as in the view of Churchill and others) that their courage was simply in overcoming their fear. His heroines and heroes were 'sustained altruists'. He admired the courage it must have taken to die for a cause. But he placed as much emphasis on the courage it must take to forgive for a cause.

We have to forgive each other. We have to forgive ourselves. In November 2015, two days after terrorists killed 130 people in attacks on Paris, Antoine Leiris wrote a powerful open letter to them on Facebook. His wife had been among those murdered.

'On Friday night, you stole the life of an exceptional being, the love of my life, the mother of my son, but you will not have my hate. I don't know who you are and I don't want to know. You are dead souls ... You want me to be scared, to see my fellow citizens through suspicious eyes, to sacrifice my freedom for security. You have failed. I will not change.'

He insisted that his baby son's happiness would also defy them: 'Because you will not have his hate either.'

The Garden of Forgiveness in Beirut (Hadiqat As-Samah) was designed to be a place of calm reflection, away from the noise of the city, and nestled between churches and mosques. Having read this chapter, you won't be surprised to learn that psychologist Alexandra Asseily has been at the heart of the project, a practical manifestation of this opportunity to heal the wounds of history. In line with Alexandra's approach to understanding our ancestors, the garden is also intended to use archaeology to highlight a sense of a shared past, predating the civil war. Sadly planning permission for the garden remains to be granted. Hate is still a useful political tool, and some of Lebanon's politicians are not yet ready to let people forgive.

But those warlords may not be alone. Perhaps we are more like them than we might realise or acknowledge. Every unforgiven trauma in our own lives, large or small, causes pain and corrosion. We need to find ways to forgive those who have harmed us. This forgiveness can extend to our ancestors or the ghosts in our families. Ultimately it must also extend to ourselves.

We can observe this superpower. Nelson Mandela knew, as he prepared for his release from jail, that 'as I walked out the door toward the gate that would lead to my freedom, if I didn't leave my bitterness and hatred behind, I'd still be in prison'. Gandhi's ability to forgive his opponents was the engine that drove the non-violent movement, led India to independence and inspired campaigns for rights and freedoms across the world. He assumed that, regardless of the regime under which people lived, they possessed a freedom of conscience, an inner capacity to make their own moral choices.

Because forgiveness is the hardest survival skill of them all, it is incredibly moving when we see it. What is the injustice or perceived injustice towards you personally that angers you most. How is it corrosive? How could you begin to feel your way to forgiveness? Or at least not to allow that gnawing resentment? What are the small moments of communication and connection that could start the healing process?

In 2020, just before the Covid lockdown started, I interviewed Palestinian doctor Izzeldin Abuelaish on stage at the first Hay Festival in Abu Dhabi. Izzeldin's story is inspiring and devastating. Living in Gaza and the first Palestinian doctor to practise in an Israeli hospital, he endured the checkpoints and grinding humiliation every day for his profession and his family, even after his wife Nadia died after a long illness. He had seen two family houses bulldozed by order of Israeli general (and later prime minister) Ariel Sharon, the second to make the street in the ramshackle refugee camp in which he lived wide enough for tanks to pass through.

But worse was to come. During Israel's bombing of Gaza in 2009 – bearing the terrifying title Operation Cast Lead – Izzeldin lost three beloved daughters and a niece in a single attack. The Goldstone Inquiry would later call the operation 'deliberately disproportionate'. Izzeldin's daughter Mayar had said that she wanted her kids to 'live in a reality where the word rocket is just another name for a space shuttle'.

She never saw that reality.

As I sat opposite Izzeldin in Abu Dhabi, I found it very hard to find the words and questions to capture all this. In the end, all I could do was to ask him to talk. He chose his words carefully, stepping through his memories like a man crossing a minefield. He described the sabra plant that grew on the land his family lost after the Israeli occupation. It is tenacious and resilient. He described how his daughters wrote their names in the sand on their last beach trip together. And kept rewriting them even when the sea washed the words away. 'I never tried to teach them resilience, only to see other people as like them, even their enemies. I learned patience at those checkpoints, even while Nadia was dying.'

We wept together on the stage. Izzeldin's wounds are raw, and will never heal. How do you heal from holding the broken and smashed bodies of three daughters? What more could I say than that we stood there in witness and solidarity?

Izzeldin sobbed. I reached across and gripped his arm, unsure how to react. His voice was quiet. 'You can never expect the pain to go,' he said. 'And you can show courage simply by remembering them. By carrying on.' He echoed Antoine Leiris. 'I can never not hate what they did to my daughters. But I can choose not to hate them.'

Time slowed. No one moved in the silent auditorium. We felt a powerful sense of empathy, but also a powerless sense that there was nothing left to say. Then Izzeldin took a deep breath, summoning up the strength as he must have to do so many times every day. A sigh heavy with loss and emotion. He turned from me and leaned towards the audience.

'It is so, so hard. But ultimately the greatest courage is to forgive.'

PART TWO

Begin It
Now

11

Education's Sliding Doors Moment

The goal of education is not to increase the amount of
knowledge but to create the possibilities for a child to
invent and discover, to create men who are capable of
doing new things.

Jean Piaget

The first thing that struck me when I visited the California
school for the kids of the tech titans was that they had a
picture of Mark Zuckerberg on the wall, where once might
have been a monarch or religious figure.

The second thing that struck me was that, for a place that
had been founded on disruption, there was so little – well –
disruption. I had gone to the school to see what those who
best understand the next wave of technology are teaching
their kids about how to handle it. I had expected it to feel
futuristic. It wasn't. Instead, young people were working
away, either in mixed year groups to solve problems or quietly
focused on creative projects. The tech was around, but it was

a tool in their hands. It was very clearly working for them. There was little hierarchy, structure and discipline.

In schools like that, arguments over the need for more social and emotional learning have already been won. Problem solving, team working, critical thinking and creativity have been prioritised over remembering things or passing exams. All this was deliberate. The head teacher explained that the parents spending their days creating tech to captivate, absorb and often replace us wanted their kids to do what the tech could not do: create, empathise, work together. And the kids were responding by taking more control of their learning.

As a result of these skills, the children of the tech titans won't work for the robots, the robots will work for them. They will be equipped with skills that are not easily automated. They will develop what it means to be human, with the technology enhancing their ability to influence the world around them, not to replace it.

So technology is already changing learning faster for the rich. We are at an inflection point for education. We face two very different scenarios.

First, and likeliest, the divide widens.

The wealthy have always sought to give their children a head start: it is hardwired into humans. Many of the best teachers will migrate to elite schools and to digital platforms, where they will be paid handsomely for TED talk classes. Alongside these elite platforms, online clubhouses will emerge for hothousing rich kids. More tribalised social networking will provide them with, well, social networks. Going beyond the twentieth-century curriculum, the wealthiest will also invest in creativity, emotional intelligence, resilience, problem solving, giving their kids a further advantage.

In this scenario those left out of education, or being taught the wrong things in the wrong ways, will fall further behind.

The result will be that we reproduce, deepen and increase social and economic inequality. The access to information that most young people have will improve, but it is knowledge that will be automated within a decade, leaving them unskilled and out of work. Half of all subject knowledge acquired during the first year of a technology degree is already outdated by graduation. Over two-thirds of young people will work in jobs that do not yet exist. With comparable previous leaps in innovation, we had decades or even centuries to adapt. This time we don't have the luxury of decades.

So the risk is that the kids who should be curing cancer, writing the new social contract or discovering the next energy source won't do those things. Instead, they will vote for authoritarians, who after a while will stop bothering to hold votes. We will face a century of massive migration, extremism, conflict and environmental destruction. 'This is the great civil rights issue of our time,' United Nations Global Education Envoy Gordon Brown told me. We risk a new digital divide, where only a few can educate their children in the right ways.

Under this scenario, many traditional universities will die a slow death, as young people migrate towards cheaper, quicker ways to get the qualification that Google or Microsoft need them to have or the route to a gig economy job: this is attractive to a twenty-year-old; perilous to a forty-year-old. Accreditation will break down, creating a free-for-all. Many people will fail to adapt to the reality of four or five careers in a lifetime, and end up unemployable.

Meanwhile, most teachers will face a tsunami of material and technology that will feel overwhelming. The pressure for results will make the role harder and less attractive. Some tech companies, seeing education as a market, will treat poorer schools as no more than a lab for experiments with new tech, often claiming a benign motive.

In this scenario, most young people on the planet continue to learn the wrong things in the wrong ways. We fail to spark the delight and magic of learning. We forcefeed kids what we ourselves learned, without recognising how different their lives will be. Education persistently focuses on academic knowledge instead of character and skills. The league tables that compare educational quality will continue to highlight the wrong things: conventional exam results.

And this leaves us in peril. On my watch as British ambassador in Lebanon, we failed to prevent the biggest humanitarian crisis of the twenty-first century, an hour away in Syria. I believe that we cannot take on the challenges facing society without understanding what happens when we fail to find creative ways to live together – we face carnage, industrial terror, great power conflict, poverty, displacement and a Petri dish for extremism. During my years in the Middle East, I saw what happens when education systems teach children that they are different to children in the school up the road or across the ocean. Syria and conflicts like it represent a failure to accept that others might think differently from us.

Perhaps Syria seems impossibly remote from the lives that most of us lead. It is, for now. But the breakdown we have seen would have seemed remote to those living there even a decade ago. Its starting point is often the sense that we have more dividing us than we have in common, a polarisation we have seen being played out in more visible ways in recent years in the West. Friends of mine have died because of it on the streets of Beirut, Nairobi, Paris and Yorkshire. All this after two world wars and a century of conflict, during which we sacrificed so much finding out how bad it can get and working out the first collective drafts of how to limit our worst reflexes as nations.

Three constituencies – despite the best of intentions – will not drive a change in this education model.

First, the tech industry itself.

Technology will bring extraordinary and unprecedented opportunities to learn, innovate and create together. Global citizens will gain greater control of their own lives, including their education. Learning will be more collaborative, digital and human. Computers and the internet replace the printed textbook. A smartboard replaces the blackboard. A block-chain 'wallet' replaces the exam certificate.

The 2020–21 lockdowns accelerated this process. The OECD's director for global education told me that he'd seen 'more social innovation, more technological innovation in education in 2020 than in the last six years'. Students and teachers in the recent past couldn't have imagined not needing to be physically present in a learning environment, or that they could learn from lecturers on the other side of the planet. Nor could they imagine the ability that tech has given us to learn not just what we want but when we want it – think of all those pupils over the years who have been forced to sit through classes at the worst time of day for their body clocks. Many students are now able to go at their own pace. When my eight-year-old wants to learn about hurricanes, he is on YouTube before I can tell him he will be studying them in his geography class in five years.

No one understands better the failure of current education systems than those riding the technological tsunami. But, faced with the vast bureaucratic and political challenges of the education sector, many tech leaders simply invest more time rewiring underprepared employees by giving them training in skills they lack, rather than working out how to spread the benefits of education of the heart and the hand. Instead of fixing the symptoms of the education crisis, we simply pay more for the treatments.

Often, when technological disruption hits education, it fails to bring educators with it. As campaigner Graham Brown-Martin has put it, 'The message is that "we've got self-driving cars, couldn't we have a self-driving classroom?" But that's not what learning is about. Learning is a deeply personal experience. So what a qualified teacher does is know that the children in their charge are individuals and have their own personalities and their own unique talents and interests that are changing constantly. The idea that we could dispense with the teacher and just give them a tablet? I don't know what that sounds like to your ears, but to mine it sounds horrific.'

Learning is set to be the next battleground for technology. We have to ensure that tech companies are fighting for the next generation, not over them.

A second constituency that is not going to drive the change we need: national governments. We cannot wait for politicians to lead us: that is surely a major lesson of this era of pandemic and polarisation. Most leaders I've worked with are finding it harder to be strategic, harder to get the trust they need to respond to the challenges they face, harder to form reliable alliances. This is partly because their education has failed them too. At a moment when we need stronger leadership, it is becoming harder to lead. We are still building the driverless car, but we seem to have achieved a driverless world.

To change education, we're often told, we have to change politics. But education is upstream of politics, upstream of diplomacy. To change politics, we have to change education.

You only have to look at how little the average classroom has changed over two hundred years, compared to, say, the average doctor's surgery, to see how hard reform is. A doctor from 1921 would struggle to recognise much in a modern surgery. A teacher from 1921 would feel that most of the world's classrooms were much more recognisable. Even in countries that have traditionally been seen as education leaders, political trends can often take education towards nationalism, rote learning and hierarchy instead of global citizenship, creative learning and networks. Even within the most developed economies, the gap between the best and worst education is alarming.

Lucy Kellaway, the journalist we met earlier who left a successful career to teach, was driven by a desire to prepare young people for life. 'What is the point of the learning itself?' she asks. 'If this is education – stuffing facts into students so they can pass exams – then maybe it's time to do something different. The point of education as currently configured is as a signalling device to universities and employers – students with the right exam scores are allowed on to the next phase of life. The children need the qualifications not to understand the world, but to make their way in it … But it seems to me that it doesn't have to be like this. It is perfectly possible that education could serve both functions – to learn useful and interesting things about the world and to get some qualifications that perform the same signalling function.'

As we have seen, much of twentieth-century education was about mastering content, often with an increasingly intense specialisation. But as knowledge expands and becomes more available, an extreme focus on remembering facts means that you risk forgetting how ideas connect and spark. What was more important to Marie Curie, the 'facts' she learned in her Warsaw school in the 1870s or how her father taught her to analyse, question and replace them? Our education systems still push young people to waste too much vital energy and brain capacity remembering things. For digital natives, it is second nature to reach for the internet. With easy access to more information than any previous generation, my students argue, why should they memorise anything that is on Wikipedia?

I recall my driving instructor asking me whether I wanted to learn to drive or to pass the driving test. We are in danger of only teaching our young people to pass the test. Yet our survival no longer depends on simply learning lots of information, a once vital skill that has already been automated.

The key to our ability to keep learning is creativity – the ability to perceive the world and reality in new and innovative ways, to find hidden patterns, to make connections between seemingly unrelated phenomena and to generate solutions. Yet creativity is slipping off school curricula across much of the world, crowded out by the demand to chase exam results and league tables.

As education opens up, it will be harder for governments to retain a monopoly over what young people learn. The reality is that the countries that change fastest will produce the global elite of the 2040s, filling the top international positions and dominating the global economy. Yet, when it comes to reform, governments fail again and again. They become distracted and impatient, blaming the educational establishment for its inability to change. They fall into the easy game of culture wars and identity politics. Meanwhile the rivets continue to pop.

James, a twelve-year-old British pupil in Abu Dhabi, described one consequence of this failure to reform. He is fortunate to be learning from great teachers in a secure and positive environment. But as part of a family of migrants, he was in his fourth education system in eight years. There has been little overlap on what is taught and assessed. 'Why are the French and British systems so different?' he asks. Part of the answer is that it suits the French and UK governments to keep them distinct. That sustains the national industries that accredit and validate education, and it makes it easier to imprint on younger people a stronger sense that they are first and foremost part of a nation, part of a system.

We can also assume that the UN institutions to which national governments have subcontracted the responsibility for global standards but not the resource cannot make change happen. Nor the NGOs, which are overwhelmed by dealing with the consequences of the current inequalities.

A third constituency that is unlikely to lead the change: universities. Higher education is packed full of brilliant individuals who are passionate about education. But too many universities continue to invest in the factory model that sees young people move up the pyramid depending on the knowledge that they can show they have learned. Mesmerised by international league tables, exam performance and the increasingly desperate pursuit of funding, administrators in many universities have even narrowed what students study. Selection is too often rooted in academic performance instead of potential. Technology is increasing the workloads of academics instead of freeing them for groundbreaking research and teaching. An industrial education model created in the nineteenth century and updated for the mass market of the twentieth century is no longer delivering for the twenty-first century.

This will affect the choices people make about their education. Before 1970 in no country did more than 10 per cent of the population go to university. By 2050, that figure is projected to be almost half in South Korea and Singapore and over a third in much of the West. Against the backdrop of political and economic uncertainty, online alternatives, and with the cost of education rising, I'm not so sure that higher education is as safe a bet as that figure suggests.

Disruption of universities is already happening. There are calls for courses to be reduced from three to two years. Code boot camps are now a $250 million a year industry for engineers, and more of their graduates find work than do those from equivalent courses at universities. By 2025, universities could lose 30 per cent of their market share to these leaner alternatives. NYU professor Scott Galloway sees the current model as financially successful but morally suspect. 'We're no longer public servants, but luxury goods who are drunk on exclusivity and brag about turning away 80 then 85 then 90 per cent of applicants.' In his view this will become unsustainable, with the lockdowns having created a moment when we wonder whether the Emperor is actually naked. 'I think the reckoning is on its way ... Harvard is now a $50,000 streaming platform.'

The pace of the transformation we are experiencing means that we don't have the luxury of time. The blurring of the lines between school, work and life during the Covid lockdowns brought home to millions that something had to change, and shone a powerful light on the extent to which our current global education system is inefficient and unfair.

Lucy is right: it doesn't have to be like this. This is not just about justice and equality. It is about survival.

12

Renaissance 2.0

I never teach my students. I aspire only to provide the
conditions in which they can learn.

Albert Einstein

The second scenario is that we change.

This does not mean simply expanding access to education,
or adding fashionable topics to the curriculum, or even evan-
gelising (or warning) about how the next shiny gadget can
change the structure of the classroom. Instead we can refocus
education on its original purpose: as the means of passing on
the best of what we have learned from our ancestors, and a set
of survival skills to our descendants.

Our ancestors left us an extraordinary bank of wisdom
about our place in the world, and the ways they found to
create and master the technologies of their times. But the
amount of information this generation of young people are
inheriting is growing at an exponential rate. Until 1900,
human knowledge doubled every century. By 1950, human

knowledge was doubling every twenty-five years. Now it is doubling every year.

As the next billion of us come online, we will witness the greatest expansion in access to knowledge since Johannes Gutenberg invented the printing press. If printing laid the basis for the Renaissance and Reformation, what will the internet do to politics and society, to how we see our place in the world?

That is not to say that we need to jettison knowledge altogether. Just to get better at working out what, among this overwhelming amount of data, we really need to know. I put that question to Roly Keating, who as head of the British Library is as close as anyone to being the custodian of the UK's knowledge. We wrestled over what was most important to place at the heart of a curriculum. My conclusion was that we all need a basic understanding of our universe: the earth, geography, climate, agriculture, natural resources, space and the solar system. We need the introduction that our ancestors would have taken for granted: biology, biodiversity, human and animal behaviour, conservation. There are practical aspects of science and technology that should be essential: energy sources, materials, machinery, computing. On society and beliefs, we should understand political systems, religions, culture and human rights. Everyone should be able to access a basic grounding in culture: art, architecture, music, dance, literature, painting, sculpture, the media, sport. Finally, we should learn about big-picture history: how civilisations develop, the origin of modern nations, nationalism and identity, attitudes to governance and current affairs.

But strikingly, from a guardian of knowledge, Roly prioritised something else: 'What is the message that you hang over the door as they arrive into that digital or physical space where we store knowledge? It's not that we're telling you what to find out, but we're trying to communicate a degree of purpose and values. We need people to understand how to use the knowledge.'

Already it is possible to start to at least imagine what the internet could do to our access to learning. Take LearnCloud, an online platform on which volunteer educators have created the largest open repository of free learning content. Its founder Canadian entrepreneur Tariq Fancy told me that 'humanity's greatest resource – brainpower – is untapped in many parts of the world. Half of school-age children lack a quality basic education. Every smartphone holds greater computing power than all of NASA in the 1960s. Every year the quality of free online content increases. If Wikipedia could leverage non-profit status to build such a formidable army of volunteer workers that it put its for-profit competitors out of business, why can't we leverage the wisdom of the crowd behind the noble goal of educating the world's least fortunate adults and children?'

Having given us this extraordinary tool for human ingenuity, Tim Berners-Lee is not satisfied that the web has even begun to deliver its potential as a launchpad for that ingenuity. Instead he wants the internet to open up curiosity for others: 'Our success will be measured by how well we foster the creativity of our children, whether people in the future have the tools to cure diseases, distinguish reliable information from propaganda or commercial chaff and build systems that support democracy and promote accountable debate.' Having originally made the web freely available, Tim hopes all humans will be able to access it to communicate, collaborate and innovate: the internet has changed the way we think about how we share what we know. He believes that information should be in the public domain unless there is a good reason for it not to be – not the other way round. 'Greater openness and accountability will make it easier for individuals to get more directly involved and leverage the web as a medium for positive change,' he says.

Wikipedia has made this imaginable. Launched by Jimmy Wales and Larry Sanger in 2001 (the name literally means 'quick encyclopedia'), the online encyclopedia is by far the internet's most popular reference tool, with over six million articles. Sixty-five million users access it every month. Like the world wide web, it is free – one of the few top websites that is not run by a company. Jimmy Wales puts it simply: 'Imagine a world in which every single person on the planet is given free access to the sum of all human knowledge. That's what we're doing.'

In this second scenario, education does not exist to manufacture cookie-cutter political (or maybe now, social media) followers or produce a generation with skills that are already able to be automated. It exists to produce generation after generation of global citizens who can thrive and coexist in a rapidly changing world.

We make learning a preparation for life, citizenship, well-being, community, not just a set of exams to decide our station in life.

We share this learning superpower with all humans, not just those currently advantaged. We make access to opportunity fairer.

People on the move carry a learning passport from country to country, showing what they have learned. The experience of distance learning during the pandemic should make it easier for us to imagine that we can create universities that are an idea not a building, that include people – for life – not keep them out.

Young people learn not just a list of the wars their country won but how we learned to coexist between those conflicts.

They do indeed become kinder, more curious, braver than us. They learn to think critically and to collaborate across borders, races and generations. We learn how to manage technology, not be managed by it. We make the tech work for the way we learn, not the other way round. We ensure that all young people have the critical thinking to filter this overwhelming amount of information, and to understand their relationship with technology.

I brought together thirty of the world's top higher education pioneers and leaders to debate how universities will have to adapt. We concluded that the universities of the future will be more accessible, as a resource for all of society, not just for a small group who study at them for three years. They will become a hub for sharing knowledge, not a refuge for hoarding it. They will offer more programmes for those who choose not to attend full time, allowing people to combine their learning with work and life. Their curricula will develop citizens of a global world, with the ability to connect ideas, environments and places, to experience failure and to solve problems. Universities will harness personalised learning and protect individual choices to ensure that students are able to maintain their autonomy and individualism. They will lead the ethical debate about the human values that we want to imprint in technology, and how we live with machines. Most important, as artificial intelligence carries out mechanical tasks, universities will cultivate creativity.

In the Renaissance 2.0 scenario, technology shifts the way we learn from being a process of passive listening to a much more interactive, participatory approach. We all learn differently. Those who prefer to use movement, art or music to help them learn will no longer be told to sit still, stop doodling or take off their headphones. That is great for our collective ability to learn. The rapid spread of more affordable smartphones on which we can create and access content, and decent internet and search engines can unlock the transmission of knowledge on an unprecedented scale. Skills and crafts – from painting to car maintenance to risk management – that once had to be transferred in person, through an apprenticeship with an expert or via a textbook, can be shared by video.

In this scenario, we also liberate teachers to teach, and recognise that they understand more than we do about how to teach well. We can already observe that as more of the sharing of basic information goes online, it should free teachers to spend more time on coaching and feedback, on helping to curate and explain all this information. The best content and delivery will rise to the top: why learn Pythagoras from a part-time biology teacher when you can watch the world's greatest maths teacher online? The availability of online resources also makes it easier to fill the gaps in our learning.

This scenario is only a prospect if we recognise that we need a whole-society approach to education: it is too important to leave only to teachers. But make the right changes now and we create the opportunity for Zeinab's moonshot and millions more. This is not rocket science.

But who will make this happen?

We will.

Every parent wants their child to thrive. But in the absence of ways to assess hand and heart, and because of the requirements for a place at the right university or company, we have too often been blinkered by the need to pass the test. Just sit at the back of any parent meeting in which a head teacher proposes spending more time on drama or art than exam preparation. Parents' doubts are rooted in genuine concerns for our children's prospects in a rapidly changing world.

Maybe the experience of home schooling during the pandemic could be the catalyst that shifts parents to a more active role as reformers of how and why we learn. The latest research from Rebecca Winthrop and her team at the Brookings Institution suggests this may already be happening. They asked parents in fourteen areas around the world to respond to the statement 'I believe that the most important purpose of school is ...'. In the past respondents might have said, 'academic: to prepare students for college or university through rigorous content knowledge across all academic subjects'. Or maybe 'economic: to prepare students with the skills and competencies needed for the workforce'.

Instead, in nine of the fourteen areas, parents most frequently identified children's socio-emotional development as the purpose of school. 'To help students gain self-knowledge, find their personal sense of purpose, and better understand their values.' In two areas, both in India, the most frequently identified purpose of school was: 'To prepare students to be good citizens who are prepared to lead their political and civic lives.' As children got older, the parents' priorities shifted from socio-emotional to academic development. But even then, parents saw that their kids need to learn differently: they wanted more group work, playing games or discussions.

As parents, many of us constantly assess our child's education. Rebecca's research showed that parents may do this instinctively as they go about their day, relying on a mix of informal or formal information, such as a comment by their child, an interaction with their child's teacher, a neighbour's child who has graduated from school and gone on to college, a report card or exam results, or the school's rankings and evaluations. But the research also indicated that parents increasingly judge the quality of education by their child's happiness and social development over their exam grades.

Parents, then, might be moving in the right direction. The progress is still slow, and teachers have plenty of battle scars from their efforts to shift the focus from school being an exam factory towards being something that prepares young people for life, work and citizenship. But Andreas Schleicher, the OECD's director for global education and the closest thing there is to a global head teacher, thinks the lockdowns may have changed this dynamic too. 'I think you're going to see a lot of teachers go back to their principal saying, Hey, I've learned to teach, coach, mentor, facilitate and evaluate in so many new ways, why can't we incorporate that in a regular schooling day? And you are going to see a lot of principals saying to the educational administration, Hey, we were able to tear down walls and engage our communities in the search for new forms of learning when you were not there, you should entrust us with greater professional autonomy and help build a more collaborative work culture.' He believes administrators may now search out and champion change, and look to find more effective approaches to use these innovations.

Sarah Brown has had almost a decade on the frontlines of the education effort, during which she has been the architect, ideas broker and driver of an ecosystem that includes the Global Business Coalition for Education, the Theirworld education charity and the global #AWorldatSchool movement. In that period, she has been involved in all the key breakthroughs on widening access to education. Her leadership role is unique in that it has combined being an insider with being an outsider, and campaigning and advocacy with research and pilot programmes; and it seeks to find linkages between sectors. I asked her how we can apply what she has learned to the effort to change education.

Sarah described her starting point as drawing the lessons from her personal experience of previous campaigns, such as civil rights, maternal mortality and Make Poverty History. In each case, a moment arrived to 'create disruption'. There was usually huge resistance at that point from the establishment and status quo, often 'a wall of opposition'.

Only a broad-based movement for change can take this opposition on. To build such a movement, you need to state the case clearly, add recruits creatively, force attention relentlessly and enlist figureheads who can amplify the message. It is vital to find and consistently uphold the moral core of the argument. For education, for example, the 'bad guy' is a lack of hope and opportunity.

For such a large issue as education, there are more specific challenges. It is harder to identify the opponent, as all governments and agencies agree, at least in theory, that education is a positive thing. The challenge is to show that education is actually the key to unlocking all the other, sometimes more attention-grabbing challenges, from health to climate change. The 'A World at School' campaign that Sarah developed was a classic example of this 'big tent' advocacy. It deliberately set big, broad objectives so that the myriad players in the sector had an overarching narrative. It consciously set out not to compete, but to help convene. As part of a small team of dedicated and experienced campaigners, Sarah then tried to engage specific groups such as business or young people, by appealing to their specific interests in education. She focused in particular on the areas where there were no effective existing umbrella groups.

In a period of virtue signalling and celebrity campaigners, one of the most distinctive features of Sarah Brown's approach is the emphasis on genuine partnerships, and a belief that ego-driven high-profile campaigns are not successful in the long term, unless they are connected to a broader platform. In her view, this is not just a question of taste. Modern campaigners have no choice but to lead from behind, and to give away power and control. Social media has accelerated this trend: success comes increasingly often when your message is adopted by others and moves away from your direct engagement. Leading from behind also means that it is easier to pass on advocacy ecosystems. This is much harder where campaigns revolve around a 'heroic leader'. Unless they are significantly funded, few foundations named after an individual gain longevity.

Sarah's example also shows the value of staying around: being what she calls 'sticky' and persevering. Much humanitarian action has become sidetracked by the latest crisis or campaign. On education there has been a need to hold governments to account over a long period. And to do the detail. Political leaders need to hear consistent messages from a wide range of stakeholders. As Sarah puts it, you can't always go for the low-hanging fruit, sometimes you have to 'reach for the high-hanging fruit and cling on'.

This is happening.

Many boot camp providers have emerged to fill the gaps in our education. Thousands of massive open online courses (MOOCs) are being offered by companies such as Coursera, edX, FutureLearn, LinkedIn and Amazon. MasterClass, an online education platform with modules taught by experts, is now valued at over $800 million. The founder of a similar learning platform, Udacity, told me that there was an increasing demand from adult learners for skills like creativity and problem solving, especially among those working in the tech sector. They can see from their day jobs where they can't compete with the machines, and where they can. And they understand the need to manage how the tech is changing our brains, because they are the ones helping it to do that. Surely it is no coincidence that communities in Silicon Valley are among the keenest to develop and maintain their mental health.

Some more enlightened countries are recognising the need to change the focus to creativity. They are not always the traditional education superpowers. South Korea tops the charts on international measures of academic performance, but has dramatically low student motivation and happiness. In Seoul, several pilots have been attempting to shift the culture from being one of academic attainment to one of skills and creativity. In an exam-driven system, an exam-free semester has been introduced into forty-two middle schools. Arts and humanities have been added to the traditional science and maths-heavy curriculum. Extracurricular activities are being expanded, with an emphasis on community service. These are now part of university entrance procedures.

Meanwhile, the city of Hiroshima has been piloting new self- and peer-assessments of skills and character, with a growing emphasis on creativity. They have shown that a focus on these areas – from teamwork to empathy to creativity – can drive up results in more traditionally academic subjects.

Our research with the Creative Industries Federation backed this up. Encouraging and making space on the curriculum (and in our lives) for creativity had the greatest multiplier effect. 'Science and arts are not an either/or,' we argued. 'Nobel laureates in the sciences are seventeen times more likely than the average scientist to be a painter, twelve times as likely to be a poet, and four times as likely to be a musician. Of course, not everyone can be a Nobel laureate ... but a narrow focus on science, technology and maths will not deliver the innovation and creative thinking we need. Arts, including crafts and design, are a vital part of the mix. Let's break down these artificial barriers and place arts with science, technology and maths, at the heart of the education system. Let's turn STEM into STEAM.'

The transformation I've described in this book might not sound revolutionary. It isn't. Yet it requires all of us to be part of changing the why, what and how of education. We all need to think of ourselves as lifelong learners and educators. This is harder than it sounds. It is easy to despair that it is all too complicated, the obstacles too daunting. We already feel overwhelmed, insecure, anxious about the future.

But that sense of fragility is why this effort matters so much.

Here are ten questions we must answer if we are to transform global learning:

1. How can we ensure talismanic education pioneers are heard?
2. How can pioneer governments generate peer pressure for change?
3. How can business support a second learning renaissance?
4. How can the global architecture of learning create the right environment for reform?
5. Where can we agree on global learning goals?
6. How can universities become part of the solution, not part of the problem?
7. How can we get parent support for passing on skills to thrive, not just passing exams?
8. How can young people move from being consumers to producers of their education?
9. How do you measure – fairly – the new skills and competences required to thrive?
10. How can technology level up educational opportunity rather than increase inequality?

What gives me most hope is that the generation currently in education have a much more intuitive understanding than any before them of how to adapt and transform their learning, and how to identify new gaps and opportunities for where technology can help improve education in the future. The renewal of education will be driven by young people themselves.

Elsa quit her school in Paris at the age of fifteen. But she is no classic dropout. When I met her, she had just aced her baccalaureate, and was writing a guide for others who want to 'hack the bac'. 'It is already happening,' she whispered over dinner in a Brussels restaurant. 'Young people are taking control of their future.'

Roly Keating of the British Library feels a similar optimism. 'People of our age tend to imagine that libraries are empty, they're not,' he says. 'What's growing is the millennial generation's use of libraries. In their home it's engineered for distraction. And yet they know that simply to survive and thrive, they need knowledge, they need the ability to concentrate. Libraries pre-dated the internet, and I fully expect them to outlast it. Because they represent something much more timeless as a human need, which is a concentrated place free to access, open to all. What's very striking about the architecture of the British Library is that we have huge bookcases, for example the one that contains George III's book collection. And people want to sit as close as they can to them, even if they're looking at a laptop. You can feel the sort of magnetic power of the book even as the symbol of recorded knowledge and the potential of it.'

I asked guru of new power Jeremy Heimans about the implications for learning, given this growing sense of agency among young people. He challenged me that 'if you want learners to drive the changes that you describe, it has to be compelling for them. What has made other movements really fly: they are personal; they are identity-based; they created a context in which people can work with each other sideways, form deep relationships with each other. So how do you get the learners to organise together. That's what's going to drive the movement.'

Generation X grew up playing Tetris. It was neat, top-down, with clear rules. Our children are playing Minecraft. It requires more collaboration, more creativity. Ideas spread fast, sideways. These young people will liberate themselves to unleash the ingenuity and creativity that they know they need to navigate the challenges ahead. They will be armed with a sense of purpose and the agency of the smartphone. Sometimes the older generation's role will be to be humble about where we have failed; to help them understand how and where to focus their energy; and often to get out of the way. Rather than moaning about their activism, we need to understand their desire to participate more freely and flexibly.

Perhaps the most important legacy of the generations who have gone before will be to give young people the space and tools to do that. Their desire to change puts us on the cusp of a great leap forward. With learners taking back control of their education, power shifts. Only if more humans learn the right things in the right way can we meet the challenge of the twenty-first century: how to create more winners from globalisation and technological change, while better protecting those left behind.

Imagine a world in which we fail to do that, and fail to unleash great waves of creativity, innovation, discovery and opportunity.

Imagine a world in which we succeed.

13

Humanifesto

Every great age is marked by innovation and daring.
By the ability to meet unprecedented problems
with intelligent solutions.

US President John F. Kennedy

Some of the scenarios for the coming decades that I have described in this book might leave you fearful and pessimistic. You are not alone. Many argue that we face a new four horsemen of the apocalypse. In place of conquest, war, famine and death, we face the twenty-first-century upgrade: technology, pandemic, war and climate crisis.

The pace at which technology is advancing offers huge opportunities to improve the world. But it does indeed risk our security, our liberty, our creativity and our humanity. We have not experienced our last pandemic. Covid-19 may be what journalist Ros Wynne-Jones has called 'a grim dress-rehearsal' for the emergencies to come. We have not experienced our last war. Meanwhile, the climate crisis seriously risks the only planet we have yet made habitable.

These overlapping threats do create genuine peril. British Astronomer Royal Sir Martin Rees has argued that the probability of humanity's extinction before 2100 is 50 per cent. American ecologist Garrett Hardin pointed out in 1968 that multiple individuals, acting independently, rationally consulting their own self-interest, will all too often deplete a shared, limited resource, even when it is not in anyone's long-term interest for this to happen. In 1949, Albert Einstein suggested that we don't know how World War Three will be fought, but World War Four would be fought with rocks.

When Zeid Ra'ad Al Hussein became the United Nations High Commissioner for Human Rights in 2014, no one expected him to become a warrior, campaigner, target, hero. Activists were aghast that a Jordanian prince had been appointed to take on the world's elites. Was this really the right figurehead for equality and human rights? But the world's powers saw a UN insider who would tread softly and carry a small stick. His diplomatic career – ambassador to the UN by thirty-six and to the US at forty-three – had been based on discretion, charm and tact.

Zeid quickly defied the expectations. Instead of platitudes, he used the position as UN high commissioner to call out a parade of 'xenophobes, populists and racists'. Donald Trump was guilty of 'state-sponsored child abuse'. The Islamic State were creating a 'harsh, mean-spirited, house of blood', while the response from Arab regimes was like 'trying to put a fire out with gasoline'. The Philippines leader Rodrigo Duterte needed 'a psychiatrist'. One ambassador told me, disapprovingly, that Zeid had 'a tendency to poke the hornets' nest'. Dictators were less diplomatic. Duterte called him the 'son of a whore', whose hair was thinning because he had no brain.

The North Koreans labelled him a 'plot-breeding scandal-mongerer'. Venezuela branded his time in office 'a resounding failure'. China recently called his investigations 'disgraceful' and voted to prevent him even speaking at the UN Security Council. One Russian ambassador dismissed him as 'provocative, kamikaze and unhinged'.

Zeid jettisoned the quiet diplomacy. 'To the intolerant, I'm a sort of global nightmare,' he told me, 'elected by all governments, yet critic of almost all. A Muslim, who is – confusingly to racists – white-skinned. The list of places where we can holiday is getting shorter. No one takes this job to win a popularity contest.'

Where did this come from? As a junior UN official in Bosnia, Zeid told me that he had been profoundly marked by the sight of the skull of a child as a trophy on a warlord's car. He is preoccupied with the Holocaust, unusual for an Arab prince. His hero is Ben Ferencz, the 101-year-old Nuremberg prosecutor. Zeid describes his 'moment of epiphany' as coming during a 1994 flight over Germany. Below them lay the sunlit city of Weimar, the 'home of the German enlightenment'. And then, through the cloud, he saw the remains of the Buchenwald concentration camp. 'Remember that in 1933 the Holocaust was unthinkable,' he said. 'Children need to learn the evil that bigotry and chauvinism can produce.' When setting up criminal courts or drafting speeches, Zeid returns repeatedly to the Nazi era. 'If we have learned anything from history, it is that scrambling only for ourselves, our ideology, or for our own kind, will scramble it all – eventually, sometimes horrifyingly so – for everyone.'

Zeid was a global citizen before the idea went in and then out of fashion. His grandmother was the Turkish painter Fahrelnissa Zeid, his mother Swedish and his Iraqi father is now Lord Chamberlain of Jordan. Zeid was also an outsider before he became an insider. An Arab who took up rugby to survive his English private school. A Hashemite prince who struggled through military service and grew a beard to appear less European in the royal court. 'To be re-elected would be to fail,' he says, because it would have meant a series of punches pulled with member states.

Zeid has been left bruised and exhausted by 'soporific international complacency'. He despairs at how fast we have banalised evil. 'Today oppression is again fashionable. Fundamental freedoms are in retreat. Shame is also in retreat. Xenophobes and racists are casting off any sense of embarrassment. The international system is a mirage. The UN has failed in the past. That's why I prefer to err on the side of speaking out, and why I'm so fearful of the consequence in human lives if we fail. International law is a good barricade on which to fall.'

He cites Primo Levi: 'Monsters exist, but more dangerous are the functionaries ready to act without asking questions.' Everyone likes to think that atrocities are committed only by the other side. 'Human rights violations are the sharp zig-zag lines of a seismograph flashing out warnings of a coming earthquake,' says Zeid. 'They are shuddering faster and higher. This resurgent malice, irresponsibility and eye-watering stupidity are like steam at high pressure being fed into the closed chamber of world events.'

Is this how the idealism of the construction of the global architecture ends? The retirement of a diplomat in a world of fake news and 3 a.m. tweets. Where the truth is traduced, the vulnerable are scapegoated and the rules to contain our worst instincts are ripped up. And somewhere a child's skull sits on a dashboard.

Zeid frequently had to act alone, and bears the scars. But his example does point the way towards a new generation of public figures who are able to rediscover an old but powerful notion of service to the public, and to a cause, as being more important than service to themselves or to their faction. As a result of his efforts, people mobilised in support of the most vulnerable, from Burma to Sudan. The intolerant were exposed, too slowly and often without sufficient penalty. But as we experience an era of intolerance and polarisation, the work of people like Zeid can be seen to be more important than ever.

Why? We looked at Nelson Mandela's ability to put on his opponent's shirt. His speech in 1964, when on trial for treason, turned the tables on oppression and injustice by exposing the unfairness of the entire South African legal system. Mandela went to prison for twenty-seven years but won the argument. The closing words of his three-hour statement are as powerful as any of the last century: 'I have fought against white domination, and I have fought against black domination. I have cherished the ideal of a democratic and free society in which all persons live together in harmony and with equal opportunities. It is an ideal which I hope to live for and to see realised. But, my Lord, if it needs be, it is an ideal for which I am prepared to die.'

Those words are now on the walls of South Africa's Constitutional Court. But they no longer seem as inevitable as they did for most of my lifetime. In a remarkable poem, 'The Second Coming', composed in 1919, William Butler Yeats described his sense of gloom after the First World War.

> Turning and turning in the widening gyre
> The falcon cannot hear the falconer;
> Things fall apart; the centre cannot hold;
> Mere anarchy is loosed upon the world,
> The blood-dimmed tide is loosed, and everywhere

> The ceremony of innocence is drowned;
> The best lack all conviction, while the worst
> Are full of passionate intensity.

It has often felt during the first two decades of this millennium that the worst are indeed full of passionate intensity, of zeal. And that the centre – a place of reason, uncertainty, debate, curiosity, coexistence – cannot hold.

I've realised that my way of viewing the world has been shaped to a significant extent by the sense that history is somehow a one-way street. That's my filter. Like many of my generation, my worldview was shaped by the events of 1989, with the fall of the Berlin Wall and the revolutions in east and west. I was fourteen, able to watch on television and absorb for the first time a feeling of living history. The momentous events that marked that year may be fading in the rearview mirror. But it is emerging as a chapter in its own right in the history books, perhaps the defining moment of the 1948–2016 era.

Quite rightly. From the upheavals that swept the Soviet Union out of Eastern Europe to the idea of a world wide web to the crushing of the Tiananmen Square demonstrations, 1989 was a year of tumultuous change. And the revolution was televised, so the images are tattooed on our collective memory. We remember 'Tank Man', standing alone before a column of armoured vehicles in Beijing, holding bags as if returning from a shopping trip. The denim and euphoria of the first protesters to breach the Berlin Wall. The pantomime villainy of Ferdinand and Imelda Marcos and the Ceauşescus.

But what was it really all about? The year 1989 had its own hot takes. That summer, the political scientist Francis Fukuyama famously concluded that we were witnessing the end of history. He was not saying events would stop, but that free-market liberal democracy had seen off autocracy, fascism, communism and authoritarianism for good. Humanity's path was set – convergence. Clearly not everyone in Beijing, Moscow, Tehran and the mountains of Afghanistan agreed. After 9/11, the journalist Fareed Zakaria would write that the fall of the Twin Towers signalled 'the end of the end of history'.

With the benefit of over three decades more hindsight than Fukuyama, it is easier to cast 1989 as a year when – accelerated by new technology and communications – humans took decisive steps towards greater dignity, and against controlling systems and unfair distribution of opportunities. And it was indeed a decisive moment in the globalisation of a consumerist model that has dominated the last thirty years: 1989 was the year the McDonald's fast food chain came to Moscow.

Of course, all of that looks more fragile after our contemporary upheavals. Thirty years on, it is easier to discern the vulnerability of the model that Fukuyama thought had won. It has proved far better than the systems it saw off at delivering security, justice and opportunity. But it has not proved good enough. As Fukuyama and others have observed since 2016, democracies can go backwards. Like the models it replaced, the post-1989 system has shown itself to be vulnerable to decadence and decay, corruption and corrosion. It has failed as yet to adjust to the digital age by making governments more accountable and citizens more powerful. It has too often been dominated by those who see their right to be superior trumping our rights to be equal. Meanwhile, much of the world continues to entrench inequality. The #MeToo and #BlackLivesMatter movements have reminded us that we all have much further to travel than some of us realised.

The year 1989 is also a reminder that we all choose our own lessons of history. As the Berlin Wall fell, a middle-ranking KGB officer called Vladimir Putin sulked at the failures of Russian leadership. An East Berlin quantum chemist named Angela Merkel traded academia for democratic politics. And a grandson to German immigrants, the heir to New York real estate tycoon Fred Trump, launched a board game based on his success.

As we face another period of change, 1989 also demonstrates that history somersaults and progress zig-zags. For every step forward – a ceasefire in Lebanon, the Soviets leaving Afghanistan, the first black chairman of the US Joint Chiefs of Staff – there were reactionary forces pushing in the opposite direction. Mandela remained in prison, the first Al-Qaeda cell began operating in New York, and the Iranian regime was offering rewards for the killing of writers. Was 1989 really a step forward? As China's premier Zhou Enlai said of a previous moment of change that the West sees as decisive, it is still too soon to say.

Most important, 1989 is a reminder that the arc of the moral universe will only bend towards justice with our help. Seen from 2049, this decade's struggles – for new ways of living together during a pandemic or protecting our freedoms online – will be as important as the breakthroughs of 1989.

The good news, though? In 1989, Fukuyama feared that the end of history would be boring: an 'endless solving of technical problems ... and satisfaction of sophisticated consumer demands'. Some might wish for that to be humanity's next era. But they will be disappointed. Whatever happens, the next three decades certainly won't be boring. Objects in the rearview mirror may be closer than they appear.

History didn't end in 1989. I've had to remove that filter. And much of the global order I have argued for preserving earlier in this book was of course crafted in a small group of capitals on the back not just of a military victory but of two centuries of superior firepower and imperial exploitation. The institutions that I believe are so vital to the global order were shaped by and in the image of a tiny fraction of humanity. That is a challenge to my worldview, but until I really understand my filter, I won't be able to make the contribution I aspire to make.

Can the centre hold? It is often said that a pessimist is an optimist armed with facts, but I think that we can be more hopeful. Humans have had some bad decades and some bad centuries. But our form suggests that resilience and adaptability are in our DNA. At these catalytic moments of change and challenge, we find a way to muddle through. The average human now lives twice as long and grows six inches taller than our great-great-grandparents, and we have access to a life that they could never have imagined. Extreme poverty has halved in the last fifteen years. While we fretted about Covid-19 in 2020 and 2021, polio was finally wiped out in Africa. We are two hundred times less likely to die in war than we were a century ago. We are becoming collectively richer, living longer, understanding the world better, and dying less of disease, poverty or violence.

Maybe an optimist can be a pessimist armed with facts. Yet we cannot take this progress for granted. We need to take back control. The Magna Carta, literally 'big charter', is now over eight hundred years old. It was the moment when the barons imposed on the British monarch an agreement for greater liberties and rights, a milestone on the way to many of the freedoms we enjoy today. Tim Berners-Lee, a Brit who might have done more for liberty than any other, argues that we need a new Magna Carta for the internet age. He is right: we need to find, rally round and hold fast to a simple expression of the balance between liberty and security; between freedom of expression and the rights of others. Technology can then become an extension of our humanity and not a replacement for it.

In Chapter 2, I looked at how we can pull together our own manifesto, as part of the effort to find and define our sense of purpose. If we do this collectively, what might an update to the Universal Declaration of Human Rights include? Are there some basic principles of common humanity around which to gather? Drawing on and updating existing ideas, here is a potential set of them.

- We are born free and equal in dignity, freedom and rights, regardless of our race, colour, gender, nationality, sexuality, language, religion, political view or economic situation.
- We all have an equal right to liberty, protection of the law, education, freedom of conscience and of opportunity. We all have an equal right to think, speak and meet.
- We are citizens of a shared world, and we have responsibilities to each other, to our planet and to future generations.
- We come from different cultures, but we share inherited values of compassion, solidarity and respect for others.
- We cherish our individual and collective ingenuity, recognising that we are works in progress.
- We work together to reduce inequality and strive to become good ancestors.
- We are courageous enough to live together, not despite our differences but because of them.
- The purpose of education is work, citizenship and life, not just productivity. We should pass on what we have ourselves learned: knowledge, skills and values. Education should last a lifetime.

'Never let the future disturb you,' wrote the Roman emperor and part-time Stoic philosopher Marcus Aurelius in his self-help book *Meditations*, 'you will meet it, if you have to, with the same weapons of reason which today arm you against the present.'

Those weapons of reason that can give us greatest hope today are our track record as humans and the opportunity of more power moving to individual citizens. Both now require us to take individual and collective responsibility for the new education model I have described, to use the new power in our hands.

I've told humanity's story as one of the gradual evolution of reason over craziness, expertise over instinct, community over tyranny, and honesty over lies. We have to learn from, marshal and pass on that inherited ingenuity, that story of the better angels of our nature. If we do that, technology can help us reinvigorate our creativity and politics. It can enable the emergence of a better way of living together. It can help us to reason together to chart a safe course.

The novelist Arundhati Roy has written beautifully about the 2020–21 pandemic as a portal. We are, she says, faced with 'a gateway between one world and the next' and the choice before us is whether we 'walk through it, dragging the carcasses of our prejudice and hatred, our avarice, our data banks and dead ideas, our dead rivers and smoky skies behind us' or whether we 'walk through lightly, with little luggage, ready to imagine another world. And ready to fight for it.' It is once again the choice of how we become better ancestors.

In the 1950s, German philosopher Ernst Bloch argued that you can find hope everywhere once you start to look. We just don't usually notice it. But maybe we got better at noticing hope during the long months of isolation, fear and estrangement in 2020 and 2021. Maybe we will look back and see the emergence of a deeper form of human solidarity, when we started to do more to protect the most vulnerable, to sing to each other from balconies, to value those heading towards danger to protect us from it, to hear the birdsong, to find our conviction, to reduce our footprint, to renew our politics, to wave to strangers, to hold the centre, to feel more grateful for the blessings we have, and to feel greater compassion for those who do not have them.

Maybe it made us kinder, more curious and braver.

That doesn't take a revolution: these survival skills are already part of us.

Rediscovering them is how we build herd immunity to the distrust, intolerance and extremism that will otherwise bring an end to our incredible human story. I started this book talking about evolution, and the ways in which humans have constantly updated our individual and collective survival skills. That now requires a more relentless focus than ever before. Survival of the fittest has often been misunderstood. As Darwin concluded, 'It's not the strongest of the species who survive, nor the most intelligent, but the one most responsive to change.'

I hope this book has helped you think about how to do that, and that it ultimately helps more of us to do it.

Onwards.

14

The Thirty-Nine Survival Steps You Can Start to Take Today

Whatever you can do, or dream you can do,
begin it. Boldness has genius, power and magic
in it. Begin it now.

Goethe

None of this is rocket science or brain surgery. What I've tried to do in this book is distil what I've observed in or heard from people who I think are role models for these survival skills.

On the way, I've been surprised and encouraged by the extent to which their advice has coalesced, and the way in which the different survival skills reinforce each other. I don't see this book as a request to change everything about our lives. You are almost certainly already using some or all of these ideas. But I do hope that it gives you some practical nudges and adjustments that will help our collective survival.

'A journey of a thousand miles begins with a single step.'

Here then, to recap the practical ideas in this book, are thirty-nine steps ...

PURPOSE

1. Write down what matters to you in life.

   For family, health, work, wealth, learning, write a sentence on each for:

   * My hope is that ...
   * One practical step towards this that I can take now is ...
   * One essential ally I have is ...
   * I will know I am making progress towards my hope when ...
   * The main challenge/obstacle within my control is ...
   * The main challenge/obstacle beyond my control – that I need to understand better – is ...

2. Share that with someone you trust. Hold yourself accountable.
3. Seek out more fascination. Which books, films and ideas most inspired you in the last year, and why aren't you getting more of them? Find something that startles you from a completely different field and try to apply it to your own.
4. Be the change. Write your own Malala speech. What would you be prepared to die for? How are you demonstrating the change you want to see? Who are your allies (and opponents)? What is the initiative you can take at school or at home?
5. Don't be the obstacle. Where did your values come from? How have they changed over time and generations? And why? Write down the three biggest

systemic advantages you have had, and how they have changed your prospects at crucial moments. It might have been the right school, the subtle advantage of gender or race at a job interview, or a word in the right ear from part of your inherited network. Imagine the experience at those crucial moments of someone who was denied those advantages. What will you do now to even the playing field?

6. Get a mentor.
7. Take a calculated risk. What is your maddest idea, and what is stopping you taking it on?
8. Turn back. In what area of your life do you need to take stock and accept it is better to change?

PRACTICAL

9. Design your curriculum. What are the gaps in your knowledge and skills and how can you fill them?
10. Keep a learning diary. What did you learn today?
11. Practise ingenuity. What brain exercises are you doing every day to practise your ingenuity and innovation? What practical problems are you setting yourself to solve?
12. Practise empathy. Take a moment in history or a story from a news website. Decide who you instinctively sympathise with. And then play the other side in the role play. Or write the opening statement for them at a peace conference. Closer to home, list five things you have in common with your adversary, opponent or the person who just pushed in front of you. Imagine their background, day, mood. Take

yourself out of the equation. What action can you take to show an understanding for their outlook? And where you need to take a stand, what is a quiet yet powerful way to do that?

13. Practise kindness. Aim for five positive interactions a day. Log them to hold yourself to account (until it becomes a reflex). When you believe in someone, tell them directly. Do something kind that no one sees. Take a few minutes before you go to bed to reflect upon your kindness that day. Think about the people you met and talked to, and how you treated each other. How well did you do? What could you do better? What did you learn?

14. Practise 360-degree analysis. Observe a famous individual or someone new you meet. Imagine being a reporter on not just what they said, but why. How and why did their background, nationality or gender influence their worldview?

15. Switch to a green energy tariff, solar energy and green investment/pension fund and tell your supplier why.

16. Calculate what you *really* need to live on each year. What are the savings you can make now to get there? What is your financial plan? Get an emergency fund.

17. Take back more control of your time. Write out your ideal day: what would it take to get closer to it?

18. Write your own risk register.

19. Learn a skill/craft that won't be automated in your working life. Prepare yourself for vocations with more creativity, caring or unpredictability.

20. Tell politicians what matters to you.

PEOPLE

21. Identify your core support team. Tell them.
22. Manage your energy. Write down a list of the people you feel most energised and excited to work with. Work more with them. Write down a list of people who leave you feeling drained and unexcited. Work less with them.
23. Be methodical about building a tribe. When joining a new club, school or community, work together to identify the key people to get to know: who are your network starters? What are your points of connection or difference?
24. Be more human: keep in touch, talk about your feelings, show vulnerability, ask for help.
25. Seek out the experts.
26. Observe role models. How are they using these survival skills? 'I believe X is brave. Here's why. And here's what I learn from that.'
27. Spend more time with older people.
28. Spend more time with younger people.
29. Practise extreme forgiveness. What is the injustice or perceived injustice towards you personally that angers you most? How is it corrosive? How could you begin to feel your way to forgiveness?
30. Practise being a good ancestor. What must you pass on? What must you not pass on?

PERSONAL

31. Do something truly carefree. Create space for serendipity. Daydream.
32. Eat more plants.
33. Exercise.
34. Get a full physical and mental health MOT.
35. Unplug.
36. Connect to nature and get some sunlight.
37. Learn to breathe deeply.
38. Take twenty seconds of insane courage.
39. Join the dots.

Good luck. Let me know how you get on. I hope that this survival guide can be a living document, constantly improved by much better ideas, experiences and role models. As you move through the cycle from observation to learning to practising to teaching, please share your thoughts on how we can all develop our humanifesto.

Acknowledgements

Only as I finish writing this book do I realise that it has taken me several decades to compile.

I'm a recovering ambassador. Before that I was a boxer, frontman, door-to-door salesman, teacher, barman, construction worker, student. At each of those points, and so many since, I was scaffolded by friends, family, strangers. I hope this book can pass on some of what they taught me.

The period that forged these ideas was my time in the Middle East. I arrived in Beirut as the UK's ambassador in 2011 – at thirty-six years old – with the Arab Spring just starting. There was a brief moment of hope and expectation. I tweeted (with more heady optimism than realism or foresight) that the most powerful weapon in the region was not the nuclear or chemical weapons that proliferated, but the smartphone. For a while, as autocratic regimes tumbled like dominoes and young people flooded the streets in a dazzling wave of energy and colour, it felt like it might just be true. They were using the technology in their hands to organise, encourage each other, expose injustices. They were moving too fast for the authoritarian regimes that sought to contain them. 'If you build a wall around our internet,' as one placard proclaimed, 'we'll build an internet around your wall.'

This book is also the result of a period of self-reflection. Partly the pandemic. More importantly, the collective self-examination that the social debates of the 2020s have stimulated. I had always thought of myself as a curious, open-minded, progressive person. I had MLK on my wall, marched for gay rights in the 1990s and funded a school in Nairobi.

And yet I noticed but didn't really reflect on the fact that there was only one black student among over a hundred students in my year at my Oxford college. Or none in my intake of twenty diplomats at the Foreign Office. I walked past the statue of Robert Clive outside the Foreign Office every day without questioning what it might represent. Instead of thinking about the fundamental injustices behind these realities, I gave money (or time) to help find the sticking plasters. For much of my life, I genuinely thought I was the one breaking glass ceilings.

Why do I mention this? We talk a lot about how education is failing millions. It is. But it is also failing many of us who are privileged to get a great formal education. And that has consequences. I want to thank the people, many of them part of this book, who helped me to understand that.

I also want to thank all those who helped get the book done. My agent Charlie Brotherstone saw the potential and urged me to make the time to put the ideas down. I was once again hugely fortunate to get the chance to work with the brilliant Arabella Pike, Jo Thompson, Iain Hunt and the team at HarperCollins. The hundreds of quiet steers, course corrections, challenges and reassurances made the work of writing and thinking much less daunting and much more exciting.

ACKNOWLEDGEMENTS

My team on the Global Learning Goals project – including Jeremy Chivers, Lorraine Charles, Angela Solomon, Sho Konno, Sally Mansour, Mario Zapata, Chris Wheeler, Tiril Rahn and Dima Boulad – drove the research that underpinned many of the ideas in the book. Rebecca Cox, a genius of delivery and planning, masterminded all that, kept me focused, and planned and executed the other projects that created space and energy for this one. I am grateful to the Porticus Foundation for their support for the Towards Global Learning Goals adventure.

Many of the ideas and concepts in the book were forged through teaching and interacting with the incredible students of NYU, the Gargash Diplomatic Academy and Hertford College, Oxford. Some will be able to play bingo with the familiar references to not leaving a song unsung; to being kind, curious and brave; to being a good ancestor; or to living a life shaped by eulogy rather than CV. Apologies for that and grateful thanks to them.

This book was shaped during a period after I left the career I loved. That this could be a time of learning, adventures and creativity was thanks to so many. But it included pivotal interventions from Sarah Brown, Bernardino Leon, Alec Ross, Randa Grob-Zakhary, John Sexton, Alastair Campbell.

Debates are at the heart of this book. Many may recognise their role in shaping the contours of those, including Helen Clark, Noura Al Kaabi, Ohoud Roumi, Marc Adelman, Rabih Abouchakra, Dubai Abulhoul, Mina Al-Oraibi, Valerie Amos, Cathy Ashton, Anthony Seldon, Sarah Zeid, Zeid Ra'ad Al Hussein, Alex Asseily, Ziyad Baroud, Christian Turner, Simon Bishop, Justin Forsyth, Kirsty McNeill, Alistair Burt, Tony Bury, John Casson, Karma Ekmekji, Tara Swart, Kim Ghattas, Nik Gowing, Filippo Grandi, Tomos Grace, Jon Luff, Katrin Bennhold, Helene Holm Pedersen, Casper Klynge, Will

Hutton, John Kampfner, Graeme Lamb, Richard Cripwell, Lubna Al Qasimi, Amanda McLoughlin, Nader Mousavizadeh, Stewart Wood, Lana Nusseibeh, Omar Ghobash, Andreas Schleicher. And many others.

More than any other work I've done, this book is not just a tribute to my extraordinary parents, but – I hope – a continuation of their pioneering work on learning and teaching.

Louise, my soul mate, is crucial to what makes me able to write all the good bits of this book, and is working on the rest. Je t'aimais, je t'aime et je t'aimerai.

I've chosen a life that maximises my moments with my boys, Charlie and Theo. Their questions, challenges and surprises animate this work, and make it all worthwhile.

Finally, thanks to Zeinab. I hope you feel this is a better effort to answer your questions.

Notes

Introduction: Kindling the Flame

6 *But as the project grew* 'Mobilising Parents and Learners', Towards Global Learning Goals, February 2019, https://secureservercdn.net/160.153.138.219/tvi.232.myftpupload.com/wp-content/uploads/2019/02/Mobilising-parents-and-learners-Towards-Global-Learning-Goals-IPAR-7-print-version.pdf

13 *they think more than previous generations* Jennifer McNulty, 'Youth Activism Is on the Rise Around the Globe, and Adults Should Pay Attention, Says Author', UC Santa Cruz Newscenter, 17 September 2019, https://news.ucsc.edu/2019/09/taft-youth.html

13 *One in six young people in the UK* 'Mental Health of Children and Young People in England, 2020: Wave 1 Follow-up to the 2017 Survey', NHS Digital, 22 October 2020, https://digital.nhs.uk/data-and-information/publications/statistical/mental-health-of-children-and-young-people-in-england/2020-wave-1-follow-up

14 *We are no longer taking for granted* 'Only a Third of Generation Y Think Their Generation Will Have Better Quality of Life Than Their Parents ...', Ipsos Mori, 11 March 2016, https://www.ipsos.com/ipsos-mori/en-uk/only-third-generation-y-think-their-generation-will-have-better-quality-life-their-parents

14 *'Instead of houses'* 'Instead of Houses, Young People Have Houseplants', *Economist*, 6 August 2018, https://www. economist.com/graphic-detail/2018/08/06/instead-of-houses-young-people-have-houseplants

14 *As they hit the job market* 'Jobs of the Future: 2025–2050', Resumeble, 21 April 2020, https://www.resumeble.com/career-advice/jobs-of-the-future-2025-2050

14 *A quarter of adults already say* 'Accelerating Workforce Reskilling for the Fourth Industrial Revolution', World Economic Forum White Paper, July 2017, http://www3. weforum.org/docs/WEF_EGW_White_Paper_Reskilling.pdf

15 *A quarter may have no full-time job* Cornelia Daheim and Ole Wintermann, '2050: The Future of Work. Findings of an International Delphi-Study of the Millennium Project', https://www.bertelsmann-stiftung.de/fileadmin/files/BSt/Publikationen/GrauePublikationen/BST_Delphi_E_03lay.pdf

16 *'The well-paying jobs will involve creativity'* N. W. Gleason, 'Introduction' in N. W. Gleason (ed.) *Higher Education in the Era of the Fourth Industrial Revolution* (Palgrave Macmillan, 2018), https://doi.org/10.1007/978-981-13-0194-0_1

17 *'a fusion of overwhelming technological breakthroughs'* Klaus Schwab, 'The Fourth Industrial Revolution: What It Means, How to Respond', World Economic Forum, January 2016, https://www.weforum.org/agenda/2016/01/the-fourth-industrial-revolution-what-it-means-and-how-to-respond/

20 *'known unknowns'* The phrase was made popular, perhaps inadvertently, by US Defense Secretary Donald Rumsfeld

20 *By 2100, we will take up* UN, World development indicators

20 *The planet will probably see a bigger temperature increase* Chi Xu, Timothy A. Kohler, Timothy M. Lenton, Jens-Christian Svenning and Martin Scheffer, 'Future of the Human Climate Niche', *PNAS*, 26 May 2020, 117 (21) 11350–11355, https://www.pnas.org/content/117/21/11350

21 *Europe's share of world GDP* 'The World in 2050', PwC Global, February 2017, https://www.pwc.com/gx/en/issues/economy/the-world-in-2050.html

21 *That proportion will overtake the young* United Nations, World Population Prospects 2014

21 *Advances in healthcare* David Amaglobeli, Hua Chai, Era Dabla-Norris, Kamil Dybczak, Mauricio Soto and Alexander F. Tieman, 'The Future of Saving: The Role of Pension System Design in an Aging World', IMF, 15 January 2019

22 *By 2050, over half of the world's population will face water scarcity* 'Water Scarcity', UN Water, https://www.unwater.org/water-facts/scarcity/

26 *'Uber, the world's largest taxi company'* Tom Goodwin, 'The Battle Is for the Customer Interface', TechCrunch, 3 March 2015, https://techcrunch.com/2015/03/03/in-the-age-of-disintermediation-the-battle-is-all-for-the-customer-interface/

Chapter 1: How to Take Back Control

30 *But* Homo sapiens *have* 'A Brief History of Forecasting', ForesightR, 6 May 2016, http://foresightr.com/2016/05/06/a-brief-history-of-forecasting/

31 *Futurist Ray Kurzweil* Alexandra Wolfe, 'Weekly Confidential: Ray Kurzweil', *Wall Street Journal*, 30 May 2014, https://www.wsj.com/articles/ray-kurzweil-looks-into-the-future-1401490952

31 *smart homes* Maeve Campbell, 'News Clip from 1989 Shows What Experts Thought Future Homes Would Be Like in 2020', Euronews.green, 7 February 2020, https://www.euronews.com/living/2020/02/06/news-clip-from-1989-shows-what-experts-thought-future-homes-would-be-like-in-2020

32 *'by the year 2018 nationalism'* Grace Hauck, '20 Predictions for 2020', *USA Today*, 22 December 2019, https://www.usatoday.com/story/news/nation/2019/12/22/2020-predictions-decades-ago-self-driving-cars-mars-voting/2594825001/

32 *'twenty-first-century humans'* Alvin Ward, '25 Weird Old Predictions Made About the 21st Century', Mental Floss, 5 February 2020, https://www.mentalfloss.com/article/616354/weird-predictions-about-21st-century

32 *'Don't ask the barber'* Maggie Fitzgerald, 'Warren Buffett: Don't Ask the Barber Whether You Need a Haircut', CNBC, 12 February 2019, https://www.cnbc.com/2019/02/12/warren-buffett-dont-ask-the-barber-whether-you-need-a-haircut.html

32 *Like the physicist John Dalton* Rachelle Oblack, '10 Famous Meteorologists', ThoughtCo., 3 July 2019, https://www.thoughtco.com/famous-meteorologists-3444421

32 *From his data* Rachelle Oblack, '10 Famous Meteorologists', ThoughtCo., 3 July 2019, https://www.thoughtco.com/famous-meteorologists-3444421

33 *'use artefacts'* Bobbie Johnson, 'The Professionals Who Predict the Future for a Living', *MIT Technology Review*, 26

February 2020, https://www.technologyreview.com/2020/02/26/905703/professionals-who-predict-the-future-for-a-living-forecasting-futurists/

33 *He told the* New Yorker Maria Konnikova, 'How People Learn to Become Resilient', *New Yorker*, 11 February 2016, https://www.newyorker.com/science/maria-konnikova/the-secret-formula-for-resilience

36 *blessing of John O'Donohue* The title of John O'Donohue's blessing is 'For the Breakup of a Relationship'.

37 *'Every time you borrow money'* Personal finance quotes, Goodreads, https://www.goodreads.com/quotes/tag/personal-finance

37 *'Too many people spend money'* Rob Berger, 'Top 100 Money Quotes of All Time', *Forbes*, 30 April 2014, https://www.forbes.com/sites/robertberger/2014/04/30/top-100-money-quotes-of-all-time/#5a71fbd84998

37 *'Once you have your human needs met'* Daniel Gilbert, *Stumbling on Happiness* (HarperPerennial, 2007)

38 *'Wealth consists'* Rob Berger, 'Top 100 Money Quotes of All Time', *Forbes*, 30 April 2014, https://www.forbes.com/sites/robertberger/2014/04/30/top-100-money-quotes-of-all-time/#5a71fbd84998

38 *How we perceive our relative wealth* Christopher J. Boyce, Gordon D. A. Brown and Simon C. Moore, 'Money and Happiness: Rank of Income, Not Income, Affects Life Satisfaction', *Psychological Science*, Vol. 21, No. 4 (April 2010), pp. 471–5, https://www.jstor.org/stable/pdf/41062232.pdf?refreqid=excelsior%3A4d7574d672bb3355fc4b6ac8c83f5263

38 *'similar others'* David Futrelle, 'Here's How Money Really Can Buy You Happiness', *Time*, 10 June 2016, https://time.com/collection/guide-to-happiness/4856954/can-money-buy-you-happiness/

38 *'the happy people didn't know'* David Futrelle, 'Here's How Money Really Can Buy You Happiness', *Time*, 10 June 2016, https://time.com/collection/guide-to-happiness/4856954/can-money-buy-you-happiness/

38 *Students rated experiences* Susan Kelley, 'To Feel Happier, Talk About Experiences, Not Things', *Cornell Chronicle*, 29 January 2013, https://news.cornell.edu/stories/2013/01/feel-happier-talk-about-experiences-not-things

38 *The most Googled financial questions* Natalia Lusinski, 'What the Most-Googled Personal Finance Questions Reveal About

How We Think About Money', Mic, 2 May 2019, https://www.mic.com/p/the-most-googled-personal-finance-questions-reveal-so-much-about-our-money-anxieties-17292478

39 *If we think of time rather than money* Cassie Mogilner, 'The Pursuit of Happiness: Time, Money, and Social Connection', *Psychological Science*, Vol. 21, No. 9 (September 2010), pp. 1348–54, https://www.jstor.org/stable/pdf/41062376.pdf?refreqid=excelsior%3Aafa71fa720a93d16343e52faa8010d8d

40 *Oxford experts found 'Adult Learning a "Permanent National Necessity", Report Finds'*, Department for Continuing Education, University of Oxford, https://www.conted.ox.ac.uk/news/new-report-finds-adult-learning-a-permanent-national-necessity

40 *Yet there is massive variation from country to country* OECD

41 *SkillsFuture has a huge hub of resources* 'From Ferment to Fusion', Towards Global Learning Goals, May 2018, https://secureservercdn.net/160.153.138.219/tvi.232.myftpupload.com/wp-content/uploads/2018/06/From-Ferment-to-Fusion-Towards-Global-Learning-Goals.pdf

41 *we need to keep learning* See also the BeLL project

42 *We all learn differently* '4 Types of Learners in Education', Advancement Courses, 12 October 2017, https://blog.advancementcourses.com/articles/4-types-of-learners-in-education/

42 *Kenyan javelin thrower Julius Yego* Matt Church, 'The Evolution of Education', https://www.mattchurch.com/talkingpoint/education-evolution

45 *Mark Fletcher has spent a lifetime* 'About the Author', Mark Fletcher, http://www.brainfriendlylearning.org/author.htm

Chapter 2: How to Be Curious

48 *The daughter of a Malaysian engineer* Charlotte Karp, 'Australia's Youngest Billionaire, 32, Doubles Her Wealth in Just Months', *Daily Mail Australia*, 23 June 2020, https://www.dailymail.co.uk/news/article-8449289/Australias-youngest-billionaire-Melanie-Perkins-DOUBLES-wealth.html

49 *study demonstrated that curiosity* Christopher Bergland, 'Curiosity: The Good, the Bad, and the Double-Edged Sword', *Psychology Today*, 4 August 2016, https://www.psychologytoday.com/us/blog/the-athletes-way/201608/curiosity-the-good-the-bad-and-the-double-edged-sword

49 *research has shown that when facing something uncertain* Christopher K. Hsee, Bowen Ruan, 'The Pandora Effect: The

Power and Peril of Curiosity', *Psychological Science*, Vol. 27, No. 5, pp. 659–66, https://journals.sagepub.com/doi/abs/10.1177/0956797616631733?journalCode=pssa

50 *at times we need to manage* Christopher Bergland, 'Curiosity: The Good, the Bad, and the Double-Edged Sword', *Psychology Today*, 4 August 2016, https://www.psychologytoday.com/us/blog/the-athletes-way/201608/curiosity-the-good-the-bad-and-the-double-edged-sword

51 *curiosity helps us to learn* Emily Campbell, 'Six Surprising Benefits of Curiosity', *Greater Good*, 24 September 2015, https://greatergood.berkeley.edu/article/item/six_surprising_benefits_of_curiosity

51 *Research by the University of California* Marianne Stenger, 'Why Curiosity Enhances Learning', Edutopia, 17 December 2014, https://www.edutopia.org/blog/why-curiosity-enhances-learning-marianne-stenger

52 *When we are curious about our work or study* J. M. Harackiewicz, K. E. Barron, J. M. Tauer and A. J. Elliot, 'Predicting Success in College: A Longitudinal Study of Achievement Goals and Ability Measures as Predictors of Interest and Performance from Freshman Year Through Graduation'. *Journal of Educational Psychology*, Vol. 94, No. 3 (2002), pp. 562–75

52 *'remain a lifelong student'* Indra Nooyi, 'Never Be Happy With What You Know', https://www.youtube.com/watch?v=24d4rfnsOxg

52 *Harvard Business School* Francesca Gino, 'The Business Case for Curiosity', *Harvard Business Review*, September–October 2018, https://hbr.org/2018/09/curiosity

52 *This makes it easier for us* Todd B. Kashdan, Ryne A. Sherman, Jessica Yarbro and David C. Funder, 'How are Curious People Viewed and How Do They Behave in Social Situations? From the Perspectives of Self, Friends, Parents, and Unacquainted Observers', *Journal of Personality*, Vol. 81, No. 2 (April 2013), pp. 142–54, https://www.ncbi.nlm.nih.gov/pmc/articles/PMC3430822/

52 *'they increased the purchase'* Christopher Bergland, 'Curiosity: The Good, the Bad, and the Double-Edged Sword', *Psychology Today*, 4 August 2016, https://www.psychologytoday.com/us/blog/the-athletes-way/201608/curiosity-the-good-the-bad-and-the-double-edged-sword

52 *We see people as warmer* Todd B. Kashdan and John E. Roberts, 'Trait and State Curiosity in the Genesis of

Intimacy: Differentiation from Related Constructs', *Journal of Social and Clinical Psychology*, Vol. 23, No. 6 (June 2005), https://guilfordjournals.com/doi/abs/10.1521/jscp.23.6.792.54800

52 *'In spite of illness'* Nir Evron, '"Interested in Big Things, and Happy in Small Ways": Curiosity in Edith Wharton', *Twentieth-Century Literature*, Vol 64, No. 1 (2018), pp. 79–100, https://read.dukeupress.edu/twentieth-century-lit/article-abstract/64/1/79/134007/Interested-in-Big-Things-and-Happy-in-Small-Ways

53 *But curiosity expands the empathy* Roman Krznaric, 'Six Habits of Highly Empathic People', *Greater Good*, 27 November 2012, https://greatergood.berkeley.edu/article/item/six_habits_of_highly_empathic_people1

59 *Imagine the motivation of the Kampala graduate* Clare Spencer, 'Five African Inventions to Look Out For in 2017', BBC News, 3 January 2017, https://www.bbc.co.uk/news/world-africa-38294998

59 *M&Ms were invented* Kristen Mulrooney, 'Tea Bags, Duct Tape and 18 Other Inventions Born in Times of Crisis', Interesting Things, 3 April 2020

63 *would top a list of cultural moments* '80 Moments That Shaped the World', British Council, https://www.britishcouncil.org/sites/default/files/80-moments-report.pdf

64 *Biographer Walter Isaacson has written* Walter Isaacson, 'The Real Leadership Lessons of Steve Jobs', *Harvard Business Review*, April 2012, https://hbr.org/2012/04/the-real-leadership-lessons-of-steve-jobs

65 *'You can't connect the dots looking forward'* Steve Jobs' 2005 Stanford University Commencement Address, https://www.youtube.com/watch?v=UF8uR6Z6KLc

66 *'all sorts of things can happen'* Stephanie Kwolek, https://www.azquotes.com/author/45405-Stephanie_Kwolek

67 *Albert Einstein* Brent Lambert, 'Read Albert Einstein's Letter to His 11-Year-Old Son on Joy, Time and the Secret to Learning Anything', FeelGuide, 15 June 2013, https://www.feelguide.com/2013/06/15/read-albert-einsteins-letter-to-his-11-year-old-son-on-joy-time-the-secret-to-learning-anything/

67 *'Families who do well'* 'LEGO Play Well Report 2018', https://www.legofoundation.com/en/learn-how/knowledge-base/lego-play-well-report-2018/

68 *Protecting play at home* D. Whitebread, D. Neale, H. Jensen, C. Liu, S. L. Solis, E. Hopkins, K. Hirsh-Pasek, J. M. Zosh

(2017), 'The Role of Play in Children's Development: A Review of the Evidence (Research Summary)', LEGO Foundation, DK, https://www.legofoundation.com/media/1065/play-types-_-development-review_web.pdf

68 *Play develops communication skills* J. M. Zosh, E. J. Hopkins, H. Jensen, C. Liu, D. Neale, K. Hirsh-Pasek, S. L. Solis and D. Whitebread (2017), 'Learning Through Play: A Review of the Evidence (White Paper)', LEGO Foundation, DK, https://www.legofoundation.com/media/1063/learning-through-play_web.pdf

69 *these are precisely the carefree activities* See for example from the LEGO Foundation: Bo Stjerne Thomsen, 'Play to Cope with Change', 23 April 2020, https://www.legofoundation.com/en/learn-how/blog/play-to-cope-with-change/; 'Why Learning Through Play Is Important', https://www.legofoundation.com/en/why-play/why-learning-through-play-is-important/; 'Characteristics of Playful Experiences', https://www.legofoundation.com/en/why-play/characteristics-of-playful-experiences/; Bo Stjerne Thomsen, 'How Technology and Play Can Positively Reform Our Education System', 29 June 2020, https://www.legofoundation.com/en/learn-how/blog/how-technology-and-play-can-positively-reform-our-education-system/; Ollie Bray, 'Innovating Pedagogy: Exploring New Forms of Teaching, Learning and Assessment', 13 February 2019, https://www.legofoundation.com/en/learn-how/blog/innovating-pedagogy-exploring-new-forms-of-teaching-learning-and-assessment/; Ollie Bray, 'How Playful Interventions Can Support High-Quality Learning in Schools', 28 June 2019, https://www.legofoundation.com/en/learn-how/blog/how-playful-interventions-can-support-high-quality-learning-in-schools/; 'Types of Play', https://www.legofoundation.com/en/learn-how/play-tips/types-of-play/; Lawrence Cohen, 'Tips for Playful Parenting', 12 December 2018, https://www.legofoundation.com/en/learn-how/blog/tips-for-playful-parenting/; 'Six Bricks Booklet', https://www.legofoundation.com/en/learn-how/knowledge-base/six-bricks-booklet/; 'Creating Creators', https://www.legofoundation.com/en/why-play/skills-for-holistic-development/creativity-matters/creativity-matters-report-series/creating-creators/

69 *But they all required the space to think* Alice Robb, 'The "Flow State": Where Creative Work Thrives', BBC Worklife, 5 February 2019, https://www.bbc.com/worklife/article/20190204-how-to-find-your-flow-state-to-be-peak-creative

70 *'I will always choose a lazy person'* 'Give the Hardest Job to the Laziest Person', https://ravenperformancegroup.com/give-hardest-job-laziest-person/

71 *The wheel would not have been invented* David Anthony, *The Horse, the Wheel, and Language: How Bronze-Age Riders from the Eurasian Steppes Shaped the Modern World* (Princeton University Press, 2007)

72 *'If a building doesn't'* Leo Babauta, '8 Creativity Lessons from a Pixar Animator', Zen Habits, https://zenhabits.net/pixar/

73 *'They stay away from questions'* Stephanie Vozza, '8 Habits of Curious People', Fast Company, 21 April 2015, https://www.fastcompany.com/3045148/8-habits-of-curious-people

75 *Evolutionary psychology suggests that men* J. Palomäki, J. Yan, D. Modic, M. Laakasuo, '"To Bluff like a Man or Fold like a Girl?" – Gender Biased Deceptive Behavior in Online Poker', *PLoS ONE*, Vol. 11, No. 7 (2016): e0157838, https://journals.plos.org/plosone/article?id=10.1371/journal.pone.0157838

75 *'Curious people aren't afraid'* Stephanie Vozza, '8 Habits of Curious People', Fast Company, 21 April 2015, https://www.fastcompany.com/3045148/8-habits-of-curious-people

Chapter 3: How to Find Purpose

75 *If we nurture this kind of curiosity* 'Future Technology: 22 Ideas About How to Change Our World', *Science Focus*, 31 August 2021, https://www.sciencefocus.com/future-technology/future-technology-22-ideas-about-to-change-our-world/

85 *Edelman* Edelman Trust Barometer, 19 January 2020, https://www.edelman.com/trustbarometer

86 *experiment by the primatologist Frans de Waal* Excerpt from Frans de Waal's TED Talk, 4 April 2013: https://www.youtube.com/watch?v=meiU6TxysCg

88 *The initial impression* Albert Mehrabian, *Silent Messages: Implicit Communication of Emotions and Attitudes* (Wadsworth, 1972)

90 *Microsoft creating the Xbox* John Rampton, 'Businesses That Took Huge Risks That Paid Off', Inc., 11 October 2016, https://www.inc.com/john-rampton/15-businesses-that-took-huge-risks-that-paid-off.html

90 *FedEx founder raising cash* Dave Hiskey, 'The Founder of Fedex Once Saved the Company by Taking Its Last $5,000 and Turning It into $32,000 by Gambling in Vegas', Today I Found Out, 2 June 2011, https://www.todayifoundout.com/

index.php/2011/06/the-founder-of-fedex-once-saved-the-company-by-taking-its-last-5000-and-turning-it-into-32000-by-gambling-in-vegas/

90 *'Hey, Jonny'* Chris Matyszczyk, 'Isaacson: Jobs Was Ingenious, But Not Necessarily Smart', CNET, 30 October 2011, https://www.cnet.com/news/isaacson-jobs-was-ingenious-but-not-necessarily-smart/

92 *In 2008, Engineers Without Borders Canada* Engineers Without Borders, http://reports.ewb.ca/

92 *In an experiment run by product designer* Peter Skillman Design: http://www.peterskillmandesign.com/about

92 *people with grit* Angela Duckworth, *Grit: The Power of Passion and Perseverance* (Vermilion, 2016): https://angeladuckworth.com/grit-book/

93 *'missions'* TULA Learning, http://www.tulaeducation.com/

93 *As students move through the missions* The Ultimate Learning Accelerator (TULA): http://www.wise-qatar.org/edhub/ultimate-learning-accelerator-tula, https://solve.mit.edu/challenges/youth-skills-the-workforce-of-the-future/solutions/632 and https://ph.theasianparent.com/after-school-program-tula-the-ultimate-learning-accelerator/

93 *'I hate risk'* Partners in Leadership, '5 Wildly Successful Entrepreneurs Reveal How Risk Taking Propelled Their Careers', Inc., 3 October 2018, https://www.inc.com/partners-in-leadership/5-wildly-successful-entrepreneurs-reveal-how-risk-taking-propelled-their-careers.html

Chapter 4: How to Find Your Voice

98 *book of advice* For more see Tom Fletcher, *The Naked Diplomat* (William Collins, 2016)

98 *Barack Obama commented* Lucy Fisher, 'Dream Big, Kid, and Give Carla a Call', *The Times*, 11 August 2013, https://www.thetimes.co.uk/article/dream-big-kid-and-give-carla-a-call-mtxkvdcsh88

107 *'Watch the ball!'* See W. Timothy Gallwey, *The Inner Game of Tennis: The Ultimate Guide to the Mental Side of Peak Performance* (Pan, 1974)

107 *Companies such as Cisco* 'Transitioning to Workforce 2020', Cisco White Paper (Cisco Public Information, 2011)

112 *Greta Thunberg* Rachel Elbaum and Elizabeth Chuck, 'Trump Appears to Mock Climate Change Activist Greta Thunberg in Tweet, and She Quietly Swipes Back', NBC News, 24 September 2019, https://www.nbcnews.com/politics/donald-

trump/trump-appears-mock-climate-change-activist-greta-thunberg-tweet-n1057981; Harry Pollard, 'Why Greta Thunberg Is an Inspiration', Exposure, 4 September 2019, https://exposure.org.uk/2019/09/why-greta-thunberg-is-an-inspiration/

Chapter 5: How to Find, Grow and Mobilise Your Community

119 *mobilise that community* You could begin with Christiana Figueres and Tom Rivett-Carnac's book *The Future We Choose* (Manilla Press, 2020). It has ten actions for citizens and is a manifesto for citizen action

120 *Polls suggest* Sunder Katwala, 'As Society Re-opens, Our Divides Need Not', CapX, 3 July 2020, https://capx.co/as-society-re-opens-our-divides-need-not/

120 *Ten million citizens* David Robinson, 'The Moment We Noticed: The Relationships Observatory and Our Learning from 100 Days of Lockdown', Relationships Project, 2020, http://relationshipsproject.org/content/uploads/2020/07/RP_Observatory-Report_web_final-compressed-1.pdf

121 *peaceful protests of the Liberian women* Sam Jones, '7 Times Activism Changed the World That You May Never Have Heard Of', Global Citizen, 2 June 2015, https://www.globalcitizen.org/en/content/7-times-activism-changed-the-world-that-you-may-ne/

122 *youngest-ever Nobel laureate* See https://www.malala.org/malalas-story

123 *a recent Fabian Society paper* Kirsty McNeill and Roger Harding, 'Counter Culture: How to Resist the Culture Wars and Build 21st Century Solidarity', Fabian Society, 6 July 2021, https://fabians.org.uk/publication/counter-culture/

Chapter 6: How to Coexist

133 *UNESCO's 1996 commission on education* International Commission on Education for the Twenty-first Century, Learning: The Treasure Within, UNESCO, 1996, http://www.unesco.org/education/pdf/15_62.pdf

134 *good online resources* See Training for Change: https://www.trainingforchange.org/tools/

135 *As empires rose and fell* For the textbook, I recommend Peter Frankopan's brilliant *Silk Roads* (Bloomsbury, 2015)

139 The Ascent of Man Tim Radford review of *The Ascent of Man* by Jacob Bronowski, *Guardian*, 15 April 2011, https://

www.theguardian.com/science/2011/apr/15/ascent-man-jacob-bronowski-review

140 *Since 1989* Max Roser, 'War and Peace', 2016. Published online at OurWorldInData.org, https://ourworldindata.org/war-and-peace

142 *'One hand'* Wesley Baines, '10 Times Kindness Changed the World', Beliefnet, https://www.beliefnet.com/inspiration/10-times-kindness-changed-the-world.aspx

143 *'It's coexistence or no existence'* See http://www.global education.org/quotations.htm

146 *Buddhist monk Thich Quang Duc* Robyn Johnson, '10 Revolutionary Acts of Courage by Ordinary People', Matador Network, 15 September 2008, https://matador network.com/bnt/10-revolutionary-acts-of-courage-by-ordinary-people/

147 *'I have seen how'* Barbara Kingsolver, from https://www.goodreads.com/author/quotes/3541.Barbara_Kingsolver

150 *Standing up for yourself* 'Test: Do You Stand Up for Yourself?', *Psychologies*, 22 March 2015, https://www.psychologies.co.uk/test-do-you-stand-yourself

150 *You can help kids* Kerry Flatley, 'The Best Way to Raise Assertive Kids Who Stand Up for Themselves', Self-Sufficient Kids, https://selfsufficientkids.com/kids-stand-up-for-themselves/

Chapter 7: How to Be Kind

152 *'I didn't let him win'* See https://www.youtube.com/watch?v=HYqcqzSVH60&feature=youtu.be

153 *'At the root of extreme niceness'* Marcia Sirota, 'The Difference Between Being Nice and Being Kind', *HuffPost*, 6 September 2011, https://www.huffpost.com/archive/ca/entry/too-nice_b_946956

154 *'If you want others to be happy'* Leo Babauta, 'A Guide to Cultivating Compassion in Your Life, with 7 Practices', Zen Habits, https://zenhabits.net/a-guide-to-cultivating-compassion-in-your-life-with-7-practices/

155 *A walk raises our mood* See World Happiness Report, https://worldhappiness.report/ed/2020/#read

155 *International Positive Education Network* See https://www.ipen-network.com

156 *It even changes the chemical balance* 'The Science of Kindness', Cedars Sinai, 13 February 2019, https://www.cedars-sinai.org/blog/science-of-kindness.html

156 *'crafted a species'* Dacher Keltner, *Born to Be Good: The Science of a Meaningful Life* (W. W. Norton, 2009)

156 *'scrutinized everything'* Charles Duhigg, 'What Google Learned from Its Quest to Build the Perfect Team', *New York Times Magazine*, 25 February 2016, https://www.nytimes.com/2016/02/28/magazine/what-google-learned-from-its-quest-to-build-the-perfect-team.html

159 *Alarmingly, empathy and the ability to see your own filter* Sara H. Konrath, Edward H. O'Brien, Courtney Hsing, 'Changes in Dispositional Empathy in American College Students Over Time: a Meta-Analysis', *Personality and Social Psychology Review* (May 2011), 15 (2): pp. 180–98, https://pubmed.ncbi.nlm.nih.gov/20688954/

159 *The film was developed* Jason Marsh, Vicki Zakrzewski, 'Four Lessons from Inside Out to Discuss with Kids', *Greater Good*, 14 July 2015, https://greatergood.berkeley.edu/article/item/four_lessons_from_inside_out_to_discuss_with_kids

160 *One in four Brits* Mental health facts and statistics are available at https://www.mind.org.uk/information-support/types-of-mental-health-problems/statistics-and-facts-about-mental-health/how-common-are-mental-health-problems/

161 *In an increasingly stressful* Mobilising parents & learners – Towards Global Learning Goals report

161 *Sousa Mendes* James Badcock, 'Portugal Finally Recognises Consul Who Saved Thousands from Holocaust', BBC News, 17 June 2020, https://www.bbc.com/news/world-europe-53006790

161 *Harriet Tubman* Elinor Evans, '5 Acts of Kindness That Changed History', History Extra, 17 February 2020, https://www.historyextra.com/period/20th-century/acts-kindness-history-examples-jane-austen-harriet-tubman-elizabeth-fry-jesse-owens-berlin-olympics-miep-gies-anne-frank/

162 *We already feel bombarded* Ben Goldacre, author of *Bad Science*, is on a mission is to protect the public from the stupid things that people write about science and their health. He shows how some 'nutritionists' blur research data to mystify diet and build people's dependence on their advice

162 *Ed Walsh* Ed Walsh, 'Bad Science: How to Learn from Science in the Media', *Science in School*, Issue 22, 22 February 2012, https://www.scienceinschool.org/article/2012/badscience-2/

163 *It helps your body fight infections* Heather Alexander, '5 Benefits of a Plant-Based Diet', MD Anderson Cancer Center,

University of Texas, November 2019, https://www.
mdanderson.org/publications/focused-on-health/5-benefits-of-
a-plant-based-diet.h20-1592991.html

163 *'chew to win'* Jackson Cole, 'Ray Parlour Reveals Strange
Secret Behind Arsenal's Success Under Arsene Wenger: "Chew
to Win"', Talksport, 19 June 2020, https://talksport.com/
football/719560/ray-parlour-secret-behind-arsenal-success-
arsene-wenger-chew/

163 *Dr Lilian Cheung* See www.savorthebook.com

164 *some form of exercise* 'Benefits of Exercise', NHS UK, https://
www.nhs.uk/live-well/exercise/exercise-health-benefits/

165 *The stats are worst among female and black students* Danice
K. Eaton, Lela R. McKnight-Eily, Richard Lowry, Geraldine
S. Perry, Letitia Presley-Cantrell, Janet B. Croft, 'Prevalence
of Insufficient, Borderline, and Optimal Hours of Sleep
Among High School Students – United States, 2007', *Journal
of Adolescent Health*, Vol. 46, No. 4 (January 2010),
pp. 399–401, https://www.jahonline.org/article/S1054-
139X%2809%2900600-4/fulltext

165 *Most young people* 'Daytime Somnolence', ScienceDirect,
https://www.sciencedirect.com/topics/medicine-and-dentistry/
daytime-somnolence

165 *keep regular sleep hours* 'Trouble Sleeping?', NHS UK,
https://www.nhs.uk/every-mind-matters/mental-health-issues/
sleep/

167 *We can get closer to nature* 'Making Peace with Nature',
Resurgence & Ecologist, https://www.resurgence.org/satish-
kumar/articles/nofa-interview.html

168 *You can do the same* For a more detailed way to track your
happiness, try the Oxford Happiness Questionnaire,
developed by psychologists Michael Argyle and Peter Hills.
This gives you a sense of your relative level of happiness, and
how it changes over time

169 *a more methodical approach to kindness* See also *Dare to Be
Kind: How Extraordinary Compassion Can Transform Our
World* (Legacy Lit, 2017), by Lizzie Velasquez, which is about
the path to self-acceptance, tolerance, love and forging a
compassionate world; and *Kindness: Change Your Life and
Make the World a Kinder Place* (Capstone, 2018), by Gill
Hasson, a more practical guide for how to be kind, including
how to recognise kindness all around you

169 *'Act with kindness'* From https://quotecatalog.com/quote/
8abvZQa

170 *'pay it forward'* Alice Johnston, '11 Small Acts of Kindness That Changed the World Forever', Culture Trip, 21 December 2017, https://theculturetrip.com/europe/articles/10-small-acts-of-kindness-that-changed-the-world/

170 *Harvard's Graduate School of Education* '5 Tips for Cultivating Empathy', Making Caring Common Project, Harvard Graduate School of Education, https://mcc.gse.harvard.edu/resources-for-families/5-tips-cultivating-empathy

170 *practise empathy* For more on learning empathy, do try the Empathy Lab: https://www.empathylab.uk/, including its 2020 read for empathy list: https://www.empathylab.uk/2020-read-for-empathy-collections

173 *many are unconscious* https://implicit.harvard.edu/implicit/takeatest.html offers a Harvard quiz that helps you identify bias

Chapter 8: How to Live with Technology

179 *AI can indeed be a tool* Sahajveer Baweja, Swapnil Singh, 'Beginning of Artificial Intelligence, End of Human Rights', LSE blog, 16 July 2020, https://blogs.lse.ac.uk/humanrights/2020/07/16/beginning-of-artificial-intelligence-end-of-human-rights/

179 *Through her Algorithmic Justice League* Chris Burt, 'Tech Giants Pressured to Follow Google in Removing Gender Labels from Computer Vision Services', Biometric Update, 2 March 2020, https://www.biometricupdate.com/202003/tech-giants-pressured-to-follow-google-in-removing-gender-labels-from-computer-vision-services

182 *Global Tech Panel* See https://eeas.europa.eu/topics/global-tech-panel/62657/global-tech-panel_en

182 *Oxford Commission on AI and Good Governance* See https://oxcaigg.oii.ox.ac.uk/about/

183 *International Principles on the Application of Human Rights to Communications Surveillance* See https://www.eff.org/files/necessaryandproportionatefinal.pdf. See also this from the Ford Foundation: https://www.fordfoundation.org/media/2541/prima-hr-tech-report.pdf

183 *'expose the impact'* 'Technology and Rights', Human Rights Watch, https://www.hrw.org/topic/technology-and-rights

187 *According to* Forbes Jeff Boss, *Forbes* magazine, September 2015

187 *'noise'* Daniel Kahneman, Olivier Sibony, Cass R. Sunstein, *Noise: A Flaw in Human Judgment* (William Collins, 2021)

188 *'The machinelike process'* Adam Elkus, 'The Hierarchy of Cringe', https://aelkus.github.io/culture/2019/10/21/hierarchy-of-cringe

189 *The hours* 'How Use of Social Media and Social Comparison Affect Mental Health', *Nursing Times*, 24 February 2020, https://www.nursingtimes.net/news/mental-health/how-use-of-social-media-and-social-comparison-affect-mental-health-24-02-2020/

189 *just having a phone in your pocket* Andrew K. Przybylski, Netta Weinstein, 'Can You Connect With Me Now? How the Presence of Mobile Communication Technology Influences Face-To-Face Conversation Quality', *Journal of Personal and Social Relationships*, Vol. 30, No. 3 (May 2013), pp. 237–46, https://journals.sagepub.com/doi/full/10.1177/026540751 2453827

189 *different platforms have different effects* 'Status of Mind', Royal Institute of Public Health report, 2017, https://www.rsph.org.uk/our-work/campaigns/status-of-mind.html

190 *'civic online reasoning'* Annabelle Timsit, 'In the Age of Fake News, Here's How Schools Are Teaching Kids to Think Like Fact-Checkers', Quartz, 12 February 2019, https://qz.com/1533747/in-the-age-of-fake-news-heres-how-schools-are-teaching-kids-to-think-like-fact-checkers/

190 *Schools in Finland* Jon Henley, 'How Finland Starts Its Fight Against Fake News in Primary Schools', *Guardian*, 29 January 2020, https://www.theguardian.com/world/2020/jan/28/fact-from-fiction-finlands-new-lessons-in-combating-fake-news

190 *There are browser extensions* 'Is Your News Feed a Bubble?', PolitEcho, http://politecho.org/

191 *examples of the steps we can take* 'How to Protect Your Privacy: 21st Century Survival Skills', https://www.youtube.com/watch?v=LnV8iU0CLpg

192 *Eighty-eight per cent of teens* Social media statistics, https://www.guardchild.com/social-media-statistics-2/

192 *problems with body image* 'The Link Between Social Media and Body Image, King University Online, 9 October 2019, https://online.king.edu/news/social-media-and-body-image/

192 *affects self-esteem* Leon Festinger, *A Theory of Social Comparison Process* (1954)

192 *Ten million new photos an hour* 'Status of Mind', Royal Institute of Public Health report, 2017, https://www.rsph.org.uk/our-work/campaigns/status-of-mind.html

193 *as in the Dutch system* Bonnie J. Rough, 'How the Dutch Do Sex Ed', *The Atlantic*, 27 August 2018, https://www.theatlantic.com/family/archive/2018/08/the-benefits-of-starting-sex-ed-at-age-4/568225/

193 *Or via Snapchat?* See https://www.net-aware.org.uk/networks/snapchat/

Chapter 9: How to Be Global

195 *Andreas Schleicher* See https://www.oecd.org/education/andreas-schleicher.htm

198 *business leaders consistently say* Frank Levy and Richard J. Murnane, *Dancing with Robots: Human Skills for Computerized Work* (Third Way, 2013)

198 *fit in well with a team* 'New Vision for Education: Unlocking the Potential of Technology', World Economic Forum, 2015, http://www3.weforum.org/docs/WEFUSA_NewVisionforEducation_Report2015.pdf

198 *adaptability, creativity and teamwork* Graham Brown-Martin, 'Education and the Fourth Industrial Revolution', keynote speech, ICERI 2018

201 *The Brookings Institution* Learning Metrics Task Force, Brookings Institution, https://www.brookings.edu/product/learning-metrics-task-force/

201 *boiled them down to six* Randa Grob-Zakhary and Jessica Hjarrand, 'To Close the Skills Gap, Start with the Learning Gap', Brookings Institution, https://www.brookings.edu/opinions/to-close-the-skills-gap-start-with-the-learning-gap/

201 *the OECD has set out more detail* 'Global Competency for an Inclusive World', OECD, 2016, https://www.oecd.org/education/Global-competency-for-an-inclusive-world.pdf

201 *worked with the Asia Society* 'OECD, Asia Society Release Framework, Practical Guide for Global Competence Education', Asia Society, https://asiasociety.org/oecd-asia-society-release-framework-practical-guide-global-competence-education

201 *'the capacity to analyse'* 'OECD, Asia Society Release Framework, Practical Guide for Global Competence Education', Asia Society, https://asiasociety.org/oecd-asia-society-release-framework-practical-guide-global-competence-education

202 *'inquirers'* IB learner profile: http://www.ibo.org/contentassets/fd82f70643ef4086b7d3f292cc214962/learner-profile-en.pdf

202 *At the heart of that* Vivien Stewart, 'Preparing Students for the 21st Century', Asia Society, http://asiasociety.org/global-cities-education-network/preparing-students-21st-century

202 *a curriculum for global citizenship* Fernando M. Reimers, Vidur Chopra, Connie K. Chung, *Empowering Global Citizens: A World Course* (CreateSpace Independent Publishing Platform, 2016)

204 *The learning passport enables* 'About the Learning Passport', Learning Passport, https://www.learningpassport.org/about-learning-passport

204 *a global core curriculum* 'Islands of Opportunity', Towards Global Learning Goals, December 2018, https://secureservercdn.net/160.153.138.219/tvi.232.myftpupload.com/wp-content/uploads/2019/01/Islands-of-Opportunity-Towards-Global-Learning-Goals-singepageversion.pdf

211 *Ken Robinson argued* National Advisory Committee on Creative and Cultural Education, *All Our Futures: Creativity, Culture and Education*, May 1999, sirkenrobinson.com/pdf/allourfutures.pdf

Chapter 10: How to Be a Good Ancestor

212 *The cancer they could develop* Alice Johnston, '11 Small Acts of Kindness That Changed the World Forever', Culture Trip, 21 December 2017, https://theculturetrip.com/europe/articles/10-small-acts-of-kindness-that-changed-the-world/

215 *another adventure* Matthew Teller, 'The Three-Month Flight Along the Nile', BBC News, 5 January 2014, https://www.bbc.com/news/magazine-25578363

215 *part of honouring and remembering* This starts with learning to grieve for them: the Collective Psychology Project's 'This Too Shall Pass' report gives us a toolkit

220 *Diplowomen* 'Diplomat Interview: Karma Ekmekji', *Diplomat*, 2 July 2019, https://diplomatmagazine.com/diplomat-interview-karma-ekmekji/

222 *Food production accounts for* Here's a nifty tool to check the carbon footprint of your diet: https://www.bbc.com/news/science-environment-46459714

223 *has replaced the trees* The Intergovernmental Panel on Climate Change, https://www.ipcc.ch/

223 *a two-year programme* EAT-Lancet Commission on Food, Plant, Health, https://eatforum.org/eat-lancet-commission/

223 *transform our understanding of inheritance and evolution* Philip Hunter has argued that epigenetic inheritance 'is

implicated in the passing down of certain cultural, personality or even psychiatric traits'. Philip Hunter, 'What Genes Remember', *Prospect*, 24 May 2008, https://www.prospectmagazine.co.uk/magazine/whatgenesremember

225 *predicted an increase in the number of conflicts* Stephen M. Walt, 'What Will 2050 Look Like?', *Foreign Policy*, 12 May 2015, https://foreignpolicy.com/2015/05/12/what-will-2050-look-like-china-nato/

229 *Hay Festival* 'Izzeldin Abuelaish Talks to Tom Fletcher', https://www.hayfestival.com/p-16413-izzeldin-abuelaish-talks-to-tom-fletcher.aspx?skinid=16

Chapter 11: Education's Sliding Doors Moment

243 *By 2050, that figure is projected to be* Max Roser and Esteban Ortiz-Ospina, 'Tertiary Education', published online at OurWorldInData.org, https://ourworldindata.org/tertiary-education

244 *'I think the reckoning'* Tim Levin, 'Education Is More Ripe for Disruption Than Nearly Any Other Industry, Says NYU Professor Scott Galloway: "Harvard Is Now a $50,000 Streaming Platform"', *Business Insider*, 9 December 2020, https://www.businessinsider.com/nyu-professor-scott-galloway-college-access-equity-future-education-disruption-2020-12

Chapter 12: Renaissance 2.0

249 *I brought together thirty* 'Universities of the Future', Towards Global Learning Goals, https://globallearninggoals.org/universities-of-the-future/

250 *shared by video* Samo Burja, 'The YouTube Revolution in Knowledge Transfer', Medium, 17 September 2019, https://medium.com/@samo.burja/the-youtube-revolution-in-knowledge-transfer-cb701f82096a

251 *'I believe that'* Rebecca Winthrop and Mahsa Ershadi, 'Know Your Parents: A Global Study of Family Beliefs, Motivations, and Sources of Information on Schooling', Brookings Institution, 16 March 2021, https://www.brookings.edu/essay/know-your-parents/

254 *MasterClass, an online education platform* See https://www.bloombergquint.com/business/masterclass-is-said-to-seek-funding-at-about-800-million-value

255 *an exam-free semester* Jung Min-ho, 'Exam-Free Semester Program Gets Positive Reviews', *Korea Times*, 9 December

2014, http://www.koreatimes.co.kr/www/news/
nation/2014/12/116_169600.html

255 *part of university entrance procedures* Vivien Stewart,
'Preparing Students for the 21st Century', Asia Society,
http://asiasociety.org/global-cities-education-network/
preparing-students-21st-century

255 *the city of Hiroshima* Vivien Stewart, 'Preparing Students for
the 21st Century', Asia Society, http://asiasociety.org/global-
cities-education-network/preparing-students-21st-century

255 *'Science and arts'* Creative Education Agenda: How and Why
the Next Government Should Support Cultural and Creative
Learning in the UK, 2017, https://www.creativeindustries
federation.com/sites/default/files/2017-05/CIF_EduAgenda_
spreads.pdf

Chapter 13: Humanifesto

269 *'a gateway between one world'* Arundhati Roy, 'The
Pandemic Is a Portal', *Financial Times*, 3 April 2020, https://
www.ft.com/content/10d8f5e8-74eb-11ea-95fe-fcd274e920ca

Chapter 14: The Thirty-Nine Survival Steps You Can Start to Take Today

271 *'A journey of a thousand miles'* Attributed to the Chinese
philosopher Lao Tzu

274 *Prepare yourself for vocations* Arwa Mahdawi, 'What Jobs
Will Still Be Around in 20 Years? Read This to Prepare Your
Future', *Guardian*, 26 June 2017, https://www.theguardian.
com/us-news/2017/jun/26/jobs-future-automation-robots-
skills-creative-health

276 *Get a full physical* Kit Collingwood, 'A Mental Health
MOT', Medium, 1 October 2019, https://medium.com/
oneteamgov/a-mental-health-mot-a1fdc6cc2225

Index